The Ancient Alien Hypothesis

A Comprehensive Guide to Extraterrestrial Influence on Human History

William M. Rowe

Rowe Publishing Company

Copyright

This publication is designed to provide accurate and authoritative information in regard to the subject matter covered. It is sold with the understanding that neither the author nor the publisher is engaged in rendering legal, investment, accounting or other professional services. While the publisher and author have used their best efforts in preparing this book, they make no representations or warranties with respect to the accuracy or completeness of the contents of this book and specifically disclaim any implied warranties of merchantability or fitness for a particular purpose. No warranty may be created or extended by sales representatives or written sales materials. The advice and strategies contained herein may not be suitable for your situation. You should consult with a professional when appropriate. Neither the publisher nor the author shall be liable for

The Ancient Alien Hypothesis: A Comprehensive Guide to Extraterrestrial Influence on Human History

UFO and Ancient Alien Artwork: Rowe Publishing Company
UFO Illustrations: Rowe Publishing Company
Cover Design: Rowe Publishing Company

Printed in The United States of America

First edition 2023

Content

Foreword

The Ancient Alien Hypothesis

In an era marked by rapid scientific advancements and an ever-growing thirst for knowledge, "The Ancient Alien Hypothesis: A Comprehensive Guide to Extraterrestrial Influence on Human History" emerges as a compelling and thought-provoking exploration of the mysteries that pervade our past. This extraordinary book delves deep into the enigmatic corners of human history, presenting a persuasive argument for the existence of ancient aliens and their role in shaping the course of our civilization. Through a careful examination of ancient artifacts, mysterious structures, and captivating myths, this volume makes a compelling case in favor of the existence of extraterrestrial beings in our past.

At the heart of this book lies the essential question that has fascinated scholars and enthusiasts alike for decades: Is there evidence of intelligent life beyond Earth? By providing a comprehensive analysis of the various theories related to the Ancient Astronaut Hypothesis, this book invites readers to embark on a fascinating intellectual journey that spans the realms of science, mythology, and spirituality. Through this exploration, readers will gain a deeper understanding of the resourcefulness and ingenuity of our ancestors and the complexities of their societies.

The chapters within this book meticulously examine the many facets of the Ancient Alien Theory, from accounts of alien encounters and hybridization to the role of governments and science in the cover-up of extraterrestrial life. The book also delves into the search for extraterrestrial communication, alien abduction experiences, and unexplained phenomena such as crop circles and animal mutilations. These intriguing topics, along with many others, serve to paint a compelling picture of a world that has been shaped by forces beyond our own.

One of the most remarkable aspects of the "The Ancient Alien Hypothesis" is its capacity to bridge the gap between scientific inquiry and mythology. By examining ancient texts through the lens of this theory, scholars have uncovered potential evidence of extraterrestrial contact in diverse cultures and religious traditions. This book delves into these ancient texts and provides a comprehensive analysis of the clues that suggest the involvement of otherworldly beings in our past.

Moreover, this book explores the future of Ancient Alien Theory, considering whether it will remain an area of research, inspire creative science fiction, or perhaps even pave the way for new technological advancements and interstellar travel. As we continue to explore the depths of space and unlock the mysteries of our own history, "The Ancient Alien Hypothesis: A Comprehensive Guide to Extraterrestrial Influence on Human History" will undoubtedly play a crucial role in shaping our understanding of the cosmos and our place within it.

In conclusion, this book offers a comprehensive and engaging exploration of the "The Ancient Alien Hypothesis", inviting readers to question established beliefs and consider the possibility of extraterrestrial involvement in our past. By presenting a compelling case in favor of the existence of ancient aliens, this book challenges us to expand our understanding of the universe and reevaluate our assumptions about the nature of human history. As we embark on this intellectual journey, we open ourselves to new possibilities and forge a deeper connection with the cosmos and the mysteries that it holds.

Chapter 1

Introduction

The Origins and History of Ancient Alien Theory

Ancient Alien Theory is a fascinating subject that has captivated the imaginations of countless individuals throughout history. It proposes that extraterrestrial beings visited Earth in the distant past, making contact with early human civilizations and influencing their development through the provision of advanced technology, guidance, and knowledge. This chapter will explore the origins and history of Ancient Alien Theory, presenting scientifically accurate information and examining the association between this theory and evidence that supports the existence of ancient aliens.

The seeds of Ancient Alien Theory can be traced back to the mythologies and religious texts of various ancient civilizations. Throughout history, stories of gods or celestial beings descending from the heavens have captivated the human imagination. These accounts, found in the cultural narratives of the Sumerians, Egyptians, Greeks, and Native

Americans, among others, have provided the foundation for the modern concept of Ancient Alien Theory.

The Sumerians, one of the earliest known civilizations, left behind a rich collection of texts and myths that detailed their interactions with the Anunnaki, a pantheon of gods who were said to have descended from the heavens. These accounts describe the Anunnaki as bestowing knowledge and technology upon the Sumerians, which enabled them to make significant advancements in agriculture, architecture, and the arts.

Similarly, ancient Egyptian mythology includes accounts of gods who came from the stars and played a crucial role in shaping human civilization. These gods, such as Osiris and Isis, were described as possessing advanced knowledge and abilities, which they shared with the ancient Egyptians, allowing them to create the monumental structures and sophisticated society for which they are known.

In Greek mythology, gods like Zeus and Poseidon were said to interact with humans, often sharing their knowledge and even fathering children with mortal women. These demigods, such as Hercules and Perseus, possessed extraordinary abilities, providing further evidence of the potential influence of extraterrestrial beings on human history.

Native American mythologies also contain stories of celestial beings, known as Star People or Sky Beings, who descended from the heavens to interact with various tribes. These beings were said to have shared their wisdom and knowledge with the indigenous people, influencing their spiritual practices and cultural development.

Although these ancient accounts laid the groundwork for the concept of ancient aliens, it wasn't until the 19th and 20th centuries that the idea began to crystallize into a distinct theory. Early proponents of the idea, such as Charles Fort and H.P. Lovecraft, wrote about the possibility of extraterrestrial influences on Earth's history, laying the foundation for future exploration of the topic.

Charles Fort, an American researcher and writer, was among the first to suggest that Earth had been visited by extraterrestrial beings in the past. His books, published in the early 20th century, chronicled numerous unexplained phenomena, which he believed were evidence of extraterrestrial involvement in human history. Fort's work inspired generations of researchers and writers to explore the concept further, paving the way for the development of the Ancient Astronaut Hypothesis.

H.P. Lovecraft, an influential writer of horror and science fiction, also contributed to the early beginnings of Ancient Alien Theory. His Cthulhu Mythos, a collection of interconnected stories, introduced the concept of ancient, powerful extraterrestrial beings who once ruled Earth and influenced human civilization. Although Lovecraft's work was fictional, it sparked the imagination of countless readers and inspired many to consider the possibility of ancient alien influences on Earth's history.

As these early proponents laid the groundwork for Ancient Alien Theory, the concept began to gain traction and capture the imaginations of people worldwide. The combination of ancient mythologies, religious texts, and the writings of influential figures like Fort and Lovecraft laid the foundation for the development and acceptance of Ancient Alien Theory as a distinct and intriguing subject in its own right.

The Ancient Astronaut Hypothesis, a cornerstone of Ancient Alien Theory, emerged in the 20th century as a result of several influential books and theories. This hypothesis proposes that extraterrestrial beings visited Earth in the distant past, interacting with ancient human civilizations and influencing their development. The hypothesis has captured the imagination of countless individuals and has been the subject of numerous books, documentaries, and television series.

Swiss author Erich von Däniken's 1968 book, "Chariots of the Gods?", was instrumental in popularizing the idea that ancient human civilizations were visited and influenced by extraterrestrial beings. Von Däniken's work examined various archaeological and historical anomalies as potential evidence of extraterrestrial intervention, sparking a renewed interest in the concept and inspiring countless others to explore the subject further.

Von Däniken presented a wide range of evidence in support of the Ancient Astronaut Hypothesis, such as the Nazca Lines in Peru, the Moai statues on Easter Island, and the Great Pyramid of Giza in Egypt. He argued that these structures and artifacts were too advanced for the ancient civilizations that created them, suggesting that they must have had assistance from technologically advanced extraterrestrial beings.

The impact of "Chariots of the Gods?" on the popularization of the Ancient Astronaut Hypothesis cannot be overstated. The book became an international bestseller, selling millions of copies and igniting a passionate debate about the possibility of ancient alien visitation. The success of the book led to a series of follow-up works by von Däniken, as well as numerous documentaries and television series that explored the subject further.

Another significant contributor to the development of the Ancient Astronaut Hypothesis was Zecharia Sitchin, who, in his Earth Chronicles series, proposed that the Anunnaki, a race of ancient Sumerian gods, were in fact extraterrestrial beings from a hypothetical planet called Nibiru. Sitchin's work, which combined elements of archaeology, ancient texts, and astronomy, helped to advance the idea that extraterrestrials played a significant role in the development of early human civilizations.

Sitchin's research focused on the interpretation of ancient Sumerian texts, particularly the creation myth known as the Enuma Elish and the Epic of Gilgamesh. He argued that these texts contained evidence of extraterrestrial visitation, with the Anunnaki being depicted as advanced beings from another planet who came to Earth in search of resources, particularly gold.

According to Sitchin, the Anunnaki created humanity as a slave race to mine gold for them, and they later shared their knowledge and technology with ancient civilizations, leading to the development of advanced societies such as Sumer, Egypt, and the Indus Valley. Sitchin's work provided a compelling narrative that connected the dots between various archaeological and historical anomalies, further fueling interest in the Ancient Astronaut Hypothesis.

The emergence of the Ancient Astronaut Hypothesis has had a profound impact on the way people view human history and the potential existence of extraterrestrial life. The hypothesis has inspired a wide range of research, with numerous authors, scholars, and researchers exploring the subject and presenting their findings in books, documentaries, and television series.

The Ancient Astronaut Hypothesis has also influenced popular culture and science fiction, with films, television shows, and novels exploring the idea of ancient alien visitation and the consequences it may have had on human history. Examples include the "Stargate" franchise, which revolves around the idea that advanced alien civilizations used a network of stargates to travel between planets and interact with ancient human cultures, and the "Prometheus" film, which explores the concept of ancient aliens as the creators of humanity.

Despite the criticisms and skepticism from mainstream academia, the Ancient Astronaut Hypothesis continues to attract a dedicated following of researchers and enthusiasts who believe that the evidence in favor of extraterrestrial intervention in our past is compelling. This growing interest has led to the establishment of organizations, conferences, and events focused on discussing and promoting the idea of ancient aliens.

As technology advances and our understanding of both human history and the universe expands, new discoveries are continually being made that add to the body of evidence supporting the Ancient Astronaut Hypothesis. For example, recent discoveries of potentially habitable exoplanets and the ongoing search for extraterrestrial life through projects like the Search for Extraterrestrial Intelligence (SETI) have lent credibility to the idea that intelligent life may exist elsewhere in the universe.

Additionally, new archaeological discoveries, such as the unearthing of previously unknown ancient sites and the reevaluation of existing artifacts and structures, continue to fuel the debate surrounding the Ancient Astronaut Hypothesis. As our understanding of the past deepens, proponents of the hypothesis argue that the likelihood of extraterrestrial involvement in the development of ancient civilizations becomes more plausible.

The emergence of the Ancient Astronaut Hypothesis in the 20th century has had a significant impact on popular culture, scientific research, and the way we view our past. The works of Erich von Däniken, Zecharia Sitchin, and other proponents of the hypothesis have inspired countless individuals to question the conventional understanding of human history and explore the possibility of ancient alien visitation. As our knowledge of the universe and our own history expands, the Ancient Astronaut Hypothesis will likely continue to generate interest and debate, challenging our understanding of what we know and what we have yet to discover.

The study of unidentified flying objects (UFOs) and the search for extraterrestrial life have played a crucial role in shaping the development and acceptance of Ancient Alien Theory. The rise in UFO sightings and encounters throughout the 20th century and beyond has reinforced the belief that Earth has been visited by extraterrestrial beings, both in ancient times and in the present day. The ongoing advancements in astronomy and astrobiology have also contributed to the growing consensus that life beyond our planet is not only possible but highly probable.

The release of previously classified government documents has revealed numerous UFO sightings and encounters reported by credible witnesses, including military personnel and commercial airline pilots. These reports often describe advanced technologies and aerial maneuvers beyond the capabilities of known human-made aircraft. The sheer volume and consistency of these reports have led many Ancient Alien Theory proponents to argue that these encounters provide compelling evidence of extraterrestrial beings visiting Earth.

The Search for Extraterrestrial Intelligence (SETI) has been an influential force in the development of Ancient Alien Theory. SETI aims to detect and study extraterrestrial intelligence through various methods, such as monitoring radio and optical signals from space. While no definitive signals have been detected thus far, the ongoing search has fueled the idea that intelligent extraterrestrial life may exist and has visited our planet in the past.

The Drake Equation, formulated by astrophysicist Frank Drake in 1961, is a probabilistic argument that estimates the number of active, communicative extraterrestrial civilizations in our galaxy. The equation takes into account factors such as the rate of star formation, the fraction of stars with planets, and the likelihood of intelligent life developing. While the actual values for these factors are still debated, the Drake Equation highlights the possibility that our galaxy could be teeming with intelligent life.

The Fermi Paradox, on the other hand, questions why, if there are so many potential civilizations in our galaxy, we have not yet observed any definitive signs of extraterrestrial life. Some proponents of Ancient Alien Theory argue that the answer to the Fermi Paradox lies in our past, with extraterrestrial beings having visited Earth and interacted with ancient civilizations.

Recent advancements in technology have significantly expanded our understanding of the universe and the potential for interstellar travel. Concepts such as the Alcubierre drive, which propose bending spacetime to achieve faster-than-light travel, have opened the door to the possibility that extraterrestrial civilizations may possess the technological capabilities to traverse vast distances in space. This notion lends credibility to the idea that ancient aliens may have visited our planet in the past, sharing their knowledge and technology with early human civilizations.

The panspermia hypothesis proposes that life on Earth may have originated from extraterrestrial sources, such as microbes carried on comets or meteorites. Recent discoveries of amino acids and other organic molecules in meteorites and interstellar dust have lent support to this hypothesis. If life on Earth did originate from elsewhere in the universe, it could strengthen the argument for the existence of ancient aliens, suggesting that extraterrestrial life may have played a role in shaping the evolution of life on our planet.

The influence of UFOlogy and modern science on Ancient Alien Theory has been substantial, providing evidence and fueling speculation that supports the idea of extraterrestrial life visiting Earth throughout history. As our understanding of the cosmos continues to expand, and our ability to detect and study distant worlds improves, it is likely that the evidence supporting Ancient Alien Theory will continue to grow, further challenging our understanding of human history and our place in the cosmos.

The discovery of numerous exoplanets, particularly those located within the habitable zones of their host stars, has significantly influenced the development of Ancient Alien Theory. These findings have led to increased speculation regarding the potential for life to exist on other planets, providing a foundation for the idea that extraterrestrial life may have visited Earth in the past. As our technology advances and we gain the ability to study the atmospheres of these exoplanets, we may find evidence of extraterrestrial life, further supporting the Ancient Alien Theory.

The emerging field of astroarchaeology, which involves the search for extraterrestrial artifacts within our solar system and beyond, has also contributed to the development of Ancient Alien Theory. Proponents of this field argue that if advanced extraterrestrial civilizations have visited our solar system, they may have left behind artifacts or other evidence of their presence. The identification of such artifacts would provide definitive proof of ancient alien contact, significantly strengthening the case for the Ancient Alien Theory.

The influence of UFOlogy and modern science on Ancient Alien Theory has been substantial, providing a wealth of evidence and fueling speculation that supports the idea of extraterrestrial life visiting Earth throughout history. As our understanding of the universe continues to expand and our ability to detect and study distant worlds improves, it is likely that the evidence supporting Ancient Alien Theory will continue to grow, further challenging our understanding of human history and our place in the cosmos. The ongoing advancements in astronomy, astrobiology, and other related fields promise to bring us closer to answering the age-old question of whether or not we are alone in the universe, and whether ancient aliens have played a role in shaping our past.

Throughout history, numerous UFO sightings and encounters have been reported worldwide. These events often share striking similarities, such as the presence of advanced technology, unexplained aerial maneuvers, and encounters with humanoid beings. The fact that some of these sightings have occurred near ancient sites or in locations with historical connections to alleged extraterrestrial activity only serves to strengthen the argument for a link between UFOs and ancient aliens.

Many UFO sightings have taken place in proximity to ancient sites, such as the pyramids of Egypt, Stonehenge in England, and the Nazca Lines in Peru. These locations are often associated with remarkable architectural and engineering achievements that some believe could not have been accomplished without advanced technology or extraterrestrial intervention.

For instance, the precise alignment of the pyramids of Giza with the stars in Orion's Belt has led some researchers to suggest that their construction may have been influenced by extraterrestrial beings with advanced knowledge of astronomy. Similarly, the massive stones used to build Stonehenge, some weighing as much as 50 tons, have prompted speculation that advanced technology or even levitation techniques were employed during its construction.

The consistency of UFO sightings and encounters across different cultures and time periods is another compelling piece of evidence supporting Ancient Alien Theory. Reports of strange lights in the sky, flying objects, and encounters with humanoid beings can be found in ancient texts, such as the Indian Vedas, Chinese historical records, and the Bible.

For example, the ancient Indian text, the Mahabharata, contains descriptions of "Vimanas," flying machines that could move in any direction and were equipped with advanced weaponry. In the Bible, the story of the prophet Ezekiel's encounter with a "wheel within a wheel" has been interpreted by some as a possible UFO sighting.

Furthermore, the consistency of UFO reports in modern times, including sightings by military personnel, pilots, and astronauts, suggests that these encounters are not merely the product of overactive imaginations or cultural myths. These contemporary reports often share similar characteristics with ancient accounts, such as descriptions of advanced technology, unexplained aerial maneuvers, and humanoid beings.

The consistency and prevalence of UFO sightings and encounters throughout history and across cultures present compelling evidence for the existence of extraterrestrial beings who have visited Earth. While skeptics argue that these accounts can be explained as natural phenomena, misinterpretations, or hoaxes, proponents of Ancient Alien Theory maintain that the sheer volume and similarity of these reports cannot be dismissed so easily.

As technology advances and our understanding of the universe expands, the possibility of discovering definitive proof of extraterrestrial life grows increasingly likely. If such evidence is found, it could lend further credence to the idea that ancient aliens have played a role in shaping human history and culture, forever altering our understanding of our past and our place in the cosmos.

The rapid progress in the fields of astronomy and astrobiology has played a crucial role in supporting the idea of ancient aliens. These advancements have not only increased our knowledge about the universe but have also provided compelling evidence that life beyond Earth is highly probable.

The discovery of thousands of exoplanets in the past few decades has significantly impacted the case for ancient aliens. With the help of powerful telescopes such as the Kepler Space Telescope and the Transiting Exoplanet Survey Satellite (TESS), astronomers have identified numerous planets located within their star's habitable zone, the region around a star where conditions might be just right for liquid water to exist on a planet's surface.

The increasing number of potentially habitable exoplanets discovered so far suggests that life-sustaining planets might be more common in the universe than previously thought. As our ability to detect and study these distant worlds improves, we may soon discover planets with environments conducive to life or even signs of life itself. This would significantly bolster the case for ancient aliens, as it would demonstrate that extraterrestrial civilizations capable of visiting Earth could exist elsewhere in the universe.

Advancements in our understanding of extremophiles have further supported the idea that life could exist elsewhere in the universe. Extremophiles are organisms capable of surviving in extreme environments, such as the deep-sea hydrothermal vents, highly acidic lakes, and even within the radioactive exclusion zone around the Chernobyl nuclear disaster site.

The resilience and adaptability of these organisms demonstrate that life can thrive in a wide range of conditions, increasing the likelihood that extraterrestrial life could exist on planets with diverse and harsh environments. This adaptability also implies that life might be more widespread than previously believed, as it is not limited to Earth-like conditions.

The field of astrobiology has made significant strides in understanding the origins and potential distribution of life in the universe. Researchers have discovered complex organic molecules, the building blocks of life, in meteorites and on other celestial bodies, such as comets and moons in our solar system. These findings suggest that the ingredients for life might be widespread throughout the cosmos, raising the possibility that life may have arisen on other planets as well.

Furthermore, the study of extremophiles on Earth has provided insights into the potential types of extraterrestrial life that might exist. Microbial life, for example, could survive in subsurface oceans on moons like Europa and Enceladus, or in the subsurface of Mars. The existence of such life would provide strong evidence for the possibility of ancient alien civilizations with the capability of interstellar travel and contact with Earth.

Scientific advancements in astronomy and astrobiology have played a crucial role in supporting the idea of ancient aliens. The discoveries of exoplanets and extremophiles, as well as our growing understanding of the origins and potential distribution of life in the universe, have significantly increased the likelihood that extraterrestrial life could exist on planets with diverse and harsh environments. As our knowledge continues to expand, it is likely that the evidence supporting Ancient Alien Theory will continue to grow, further challenging our understanding of human history and our place in the cosmos.

The confluence of UFO sightings and encounters, as well as scientific advancements in the fields of astronomy and astrobiology, has had a profound impact on the development and acceptance of Ancient Alien Theory. These findings not only offer evidence for the possibility of life beyond our planet but also provoke questions about the nature of extraterrestrial life and its potential to interact with Earth.

If life exists on other planets, it is reasonable to assume that some of these lifeforms may have developed advanced civilizations with the capability to explore the cosmos. Technological advancements, such as propulsion systems that could potentially surpass the limitations of our current understanding of physics, might enable these civilizations to embark on interstellar travels. These advanced beings could have discovered Earth during their cosmic voyages, making contact with our ancestors and leaving lasting impressions on human history and culture.

The possibility of extraterrestrial contact raises questions about the influence of these beings on the development of human civilization. Proponents of Ancient Alien Theory argue that encounters with advanced extraterrestrial life could have sparked significant leaps in human knowledge, technology, and culture. These ancient aliens might have shared their wisdom, directly or indirectly, with our ancestors, shaping the course of human history and leading to the rise of great civilizations.

The ongoing prevalence of UFO sightings and encounters throughout history, up to the present day, lends credence to the idea that Earth has been, and perhaps continues to be, visited by extraterrestrial beings. These sightings often involve advanced technology, such as unexplained aerial maneuvers or objects that defy our current understanding of physics. The consistency of these reports across different cultures and time periods supports the notion of an enduring relationship between humans and extraterrestrial beings, providing further evidence for Ancient Alien Theory.

Modern scientific endeavors, such as the search for extraterrestrial life through the detection of biosignatures in the atmospheres of exoplanets, hold the potential to provide conclusive evidence for the existence of alien life. The discovery of extraterrestrial life would have significant implications for Ancient Alien Theory, as it would establish the possibility of intelligent beings capable of visiting Earth and interacting with our ancestors.

The impact of UFOlogy and modern science on Ancient Alien Theory extends beyond astronomy and astrobiology, encompassing various other disciplines such as archaeology, anthropology, and linguistics. This interdisciplinary approach enables researchers to examine the evidence from multiple perspectives, resulting in a more comprehensive understanding of the potential interactions between ancient humans and extraterrestrial beings.

The impact of UFOlogy and modern science on Ancient Alien Theory has been substantial, providing evidence for the possibility of life beyond Earth and raising questions about the nature of extraterrestrial life and its potential interactions with our planet. As our understanding of the universe continues to grow and our ability to detect and study distant worlds improves, it is likely that the evidence supporting Ancient Alien Theory will continue to accumulate, further challenging our understanding of human history and our place in the cosmos.

The Great Pyramids of Giza, located on the outskirts of modern-day Cairo, Egypt, stand as a testament to the incredible architectural and engineering achievements of the ancient Egyptians. These structures, built around 2500 BCE, continue to captivate researchers and enthusiasts alike with their impressive size, precise construction, and alignment with celestial bodies. The potential involvement of ancient aliens in the design and construction of these monuments has become a subject of great interest among proponents of the Ancient Alien Theory.

The Great Pyramid of Giza, the largest and oldest of the three pyramids, exhibits a remarkable degree of accuracy in its alignment with true north, deviating by only a fraction of a degree. In addition, the layout of the three pyramids closely mirrors the positions of the three stars that form the "belt" of the constellation Orion, which was associated with the god Osiris in ancient Egyptian mythology. This precise alignment suggests a deep understanding of astronomy and raises questions about how such knowledge was acquired at the time. Proponents of the Ancient Alien Theory argue that this level of precision and astronomical knowledge may be indicative of extraterrestrial guidance or involvement.

The construction of the pyramids required the use of massive stone blocks, some weighing as much as 70 tons. The quarries from which these stones were sourced are located several miles away from the Giza Plateau, making the transportation and placement of these enormous blocks a considerable challenge for the ancient Egyptians. Theories about how the blocks were moved range from the use of ramps and levers to the employment of vast labor forces. However, none of these theories fully explain the efficiency and precision with which the pyramids were constructed.

Furthermore, the stones used in the construction of the pyramids were cut and placed with extraordinary precision, creating joints so tight that even a sheet of paper cannot be inserted between the blocks. This level of craftsmanship is difficult to achieve even with modern technology, leading some researchers to question whether the ancient Egyptians could have accomplished this feat without external assistance.

The combination of the pyramids' precise alignment with celestial bodies, their advanced construction techniques, and the mystery surrounding the transportation and assembly of the massive stone blocks has led proponents of Ancient Alien Theory to propose that extraterrestrial beings may have played a role in the construction of the Great Pyramids of Giza.

Some researchers suggest that ancient aliens may have provided the ancient Egyptians with advanced technology or knowledge that enabled them to construct the pyramids with such precision and efficiency. Others speculate that the pyramids themselves may have served as some form of communication or energy device for extraterrestrial visitors.

While conventional explanations for the construction and purpose of the pyramids continue to be debated, the Ancient Alien Theory offers an intriguing alternative perspective on these enigmatic structures. By considering the possibility of ancient alien involvement, we can gain a deeper appreciation for the mysteries surrounding the Great Pyramids of Giza and their place in human history.

Located at an altitude of approximately 12,800 feet in modern-day Bolivia, Puma Punku is an enigmatic archaeological site that is part of the larger Tiwanaku complex, an ancient Andean civilization believed to have flourished between 500 and 1000 CE. The site features intricately carved stones with precision-cut interlocking joints, demonstrating an advanced level of engineering and construction techniques that have puzzled researchers for decades.

The most striking feature of Puma Punku is the precision with which the stones have been carved and assembled. Many of the stones display complex geometrical patterns and perfectly flat surfaces, with sharp 90-degree angles and interlocking joints that fit together seamlessly. The level of precision exhibited in these stones is remarkable, particularly considering the tools and techniques that would have been available to the ancient Tiwanaku civilization. The level of skill and craftsmanship required to achieve this feat has led proponents of Ancient Alien Theory to suggest that extraterrestrial beings may have provided the knowledge and technology necessary to accomplish such intricate work.

Another baffling aspect of Puma Punku is the transportation and assembly of the massive stones used in its construction. Some of the stones, such as the famous H-blocks, weigh over 100 tons and were quarried several miles away from the site. The difficulty of moving these enormous stones across vast distances and rough terrain, without the use of modern machinery or even the wheel, has confounded archaeologists and engineers.

Proponents of Ancient Alien Theory argue that the transportation and assembly of these massive stones could have been facilitated by advanced extraterrestrial technology, such as anti-gravity devices or other sophisticated methods unknown to us. They point to the fact that similar feats of moving and assembling colossal stones can be found in other ancient sites around the world, such as the Egyptian pyramids and Baalbek, suggesting a possible connection between these ancient civilizations and extraterrestrial visitors.

Another intriguing aspect of the Tiwanaku civilization is its apparent knowledge of astronomy. The site of Puma Punku, along with other structures in the Tiwanaku complex, appears to have been designed with astronomical alignments in mind. For example, the main temple at Tiwanaku, known as the Kalasasaya, is aligned with the cardinal directions, and its entrance is precisely oriented toward the sunrise on the December solstice.

This advanced knowledge of astronomy, combined with the enigmatic stone carvings and the transportation of massive stones, has led proponents of Ancient Alien Theory to suggest that extraterrestrial beings may have played a role in the development of the Tiwanaku civilization, imparting their knowledge of the cosmos and advanced construction techniques to the people of the region.

The remarkable features of Puma Punku and the Temple of Tiwanaku have fueled speculation that advanced extraterrestrial beings may have shared their knowledge and technology with the ancient inhabitants of the region. While conventional explanations may exist for some of these phenomena, the Ancient Alien Theory offers an alternative perspective that challenges our understanding of human history and our place in the cosmos.

The ancient site of Baalbek, located in Lebanon's Beqaa Valley, is home to some of the most enigmatic stone structures in the world. It is a testament to the architectural and engineering prowess of the civilizations that inhabited the region thousands of years ago. The foundation of the Temple of Jupiter, in particular, presents intriguing mysteries that have led some to propose that ancient aliens may have played a role in its construction.

The Trilithon is a remarkable architectural feature found at the Temple of Jupiter's foundation. It consists of three colossal limestone blocks, each weighing approximately 800 tons, measuring roughly 22 meters in length, and 4.5 meters in height. What makes these massive stones even more fascinating is their precise placement and alignment, which raises questions about the techniques used to transport and maneuver them into position.

The quarry from which these stones were sourced is located over half a mile away from the construction site. To date, researchers have been unable to determine how the ancient builders managed to transport and place the massive blocks with such precision, given the limited technology available at the time.

Some theories suggest that the blocks were moved using a combination of levers, rollers, and a vast labor force. However, the sheer size and weight of the stones make these conventional explanations difficult to accept. Proponents of Ancient Alien Theory argue that extraterrestrial beings may have provided advanced technology or guidance to the builders, enabling them to accomplish this seemingly impossible task.

The precision and craftsmanship displayed in the construction of Baalbek's stone structures are extraordinary, even by today's standards. The stones are cut and shaped with remarkable accuracy, fitting together seamlessly without the use of mortar. This level of skill and engineering is not only present in the Trilithon but also in other structures within the complex, such as the six massive columns that once supported the Temple of Jupiter's roof.

The advanced engineering techniques exhibited at Baalbek have led some to question whether human ingenuity alone was responsible for these incredible achievements or if ancient alien technology played a role in their creation.

The mysteries surrounding the construction of Baalbek's stone structures have fueled speculation that advanced extraterrestrial beings may have been involved in their creation. Ancient Alien Theory proponents argue that the incredible engineering, combined with the seemingly impossible task of transporting and placing the massive stones, points to the possibility of alien intervention.

While there is no direct evidence to support the idea of ancient alien involvement at Baalbek, the site's architectural wonders continue to defy conventional explanations. As a result, the theory that extraterrestrial beings may have provided advanced technology or guidance to the builders remains an intriguing and compelling alternative perspective.

The remarkable stone structures of Baalbek present a fascinating and enduring mystery, challenging our understanding of human capabilities in the ancient world. The enigmatic Trilithon, the precise engineering techniques, and the unknown methods of transporting the massive stones have led some to entertain the possibility of ancient alien involvement. Until definitive answers are found, Baalbek will continue to captivate the imaginations of researchers and enthusiasts alike, inspiring further exploration into the mysteries of our past and the potential influence of extraterrestrial life on human history.

The Nazca Lines are a series of massive geoglyphs etched into the desert of southern Peru, covering an area of approximately 450 square kilometers. These enigmatic designs, which include representations of animals, plants, and geometric shapes, were created between 500 BCE and 500 CE by the ancient Nazca civilization. The scale and complexity of these glyphs are best appreciated from the air, as they are barely visible from the ground, which has led many to question their purpose and the method by which they were created.

The Nazca Lines display an impressive degree of precision and scale, with some of the largest designs extending over 1,200 feet in length. The creation of these geoglyphs involved the removal of reddish-brown iron oxide-coated pebbles from the surface to reveal the lighter-colored earth underneath. This process required careful planning, precise measurements, and a remarkable understanding of the final design, which would only be visible from a considerable height.

The purpose and method of creating the Nazca Lines remain a mystery. Some researchers have suggested that they may have served a religious or ritualistic function, while others believe they were created for astronomical or calendrical purposes. However, the fact that these enormous glyphs are best viewed from the air has prompted proponents of Ancient Alien Theory to speculate that they may have been designed as messages, navigational aids, or landing markers for extraterrestrial visitors.

The extraterrestrial hypothesis proposes that the Nazca Lines were created with the assistance or guidance of advanced beings from other worlds. This theory suggests that the Nazca people may have been in contact with extraterrestrial beings who provided them with the knowledge and technology required to create these massive, intricate designs. The scale and complexity of the Nazca Lines, coupled with their visibility from the air, support the idea that they were intended to be seen by beings capable of flight or aerial observation.

The fact that the Nazca Lines are best appreciated from an aerial perspective raises questions about the knowledge and technology possessed by the Nazca civilization. How could they have created such large-scale designs without the ability to view them from above? Proponents of Ancient Alien Theory argue that this could be evidence of extraterrestrial involvement, suggesting that the Nazca people may have had access to advanced technology, such as flying vehicles or other means of aerial observation, which allowed them to plan and execute these impressive designs.

The Nazca Lines, with their enigmatic designs and impressive scale, continue to captivate researchers and enthusiasts alike. While conventional explanations for their purpose and creation exist, the Ancient Alien Theory offers a compelling alternative perspective, suggesting that these massive geoglyphs may be evidence of extraterrestrial contact and influence in our past. As we continue to explore the mysteries of our history and the universe, the Nazca Lines will remain a fascinating example of the potential impact of ancient aliens on human civilization.

The Antikythera mechanism is a remarkable ancient artifact that has continued to perplex researchers since its discovery in 1901. This device, often referred to as the world's first analog computer, is believed to have been created around 100 BCE. Its level of sophistication and precision was far ahead of its time, leading some to propose that it may be evidence of ancient alien knowledge or technology.

The Antikythera mechanism consists of a complex series of gears, dials, and pointers that work together to model the movements of celestial bodies and predict astronomical events such as eclipses. The device is housed within a wooden frame and features a sophisticated system of interlocking gears, with some estimates suggesting that it contained at least 30 individual gears. This level of complexity was unprecedented in the ancient world and would not be seen again until the development of mechanical clocks in the 14th century CE.

One of the most striking aspects of the Antikythera mechanism is its accurate representation of astronomical knowledge. The device's intricate gear system allowed it to predict the positions of the sun, moon, and possibly even the known planets with remarkable accuracy. Additionally, it incorporated advanced knowledge of lunar cycles and the Metonic cycle, a 19-year period used to synchronize solar and lunar calendars.

The level of astronomical knowledge and the precision of the calculations displayed by the Antikythera mechanism far surpass what was known to be possible in ancient Greece. This discrepancy has led some researchers to question whether the device could be the product of extraterrestrial intervention, suggesting that its creators may have been inspired by or directly assisted by advanced alien beings.

The origins of the Antikythera mechanism remain a mystery. No comparable devices have been found from the same time period, and no ancient texts or documents detail the construction or use of such a complex instrument. This lack of context has fueled speculation that the Antikythera mechanism may have been a unique artifact, possibly inspired or influenced by extraterrestrial knowledge.

The Antikythera mechanism presents a compelling case for the potential involvement of ancient aliens in human history. Its advanced design, precise calculations, and unparalleled representation of astronomical knowledge suggest that it may have been the product of a civilization with access to knowledge and technology far beyond what was known in ancient Greece.

While conventional explanations for the Antikythera mechanism's origins and construction may exist, the Ancient Alien Theory offers an intriguing alternative perspective. By considering the possibility that advanced extraterrestrial beings may have shared their knowledge and technology with ancient human civilizations, we can gain a deeper appreciation for the mysteries of our past and the potential connections between our world and the cosmos.

The analysis of ancient texts and mythology has played a significant role in the development of Ancient Alien Theory. Many ancient stories describe encounters with celestial beings or gods who descended from the heavens and shared their wisdom with humanity. These accounts can be found in texts such as the Sumerian Epic of Gilgamesh, the Hindu Vedas, the Hebrew Bible, and various Native American legends. Proponents of Ancient Alien Theory interpret these accounts as evidence of extraterrestrial visitation, pointing to the striking similarities between descriptions of celestial beings across diverse cultures and time periods.

The Sumerian Epic of Gilgamesh, one of the oldest known literary works, tells the story of a demigod, Gilgamesh, and his quest for immortality. Throughout his journey, Gilgamesh encounters divine beings and experiences extraordinary events that some proponents of Ancient Alien Theory argue may be interpreted as evidence of extraterrestrial contact. For example, the Anunnaki, a pantheon of Sumerian gods, are said to have descended from the heavens and bestowed knowledge upon humanity. Some researchers suggest that these deities were, in fact, extraterrestrial beings who visited Earth and influenced the development of ancient Sumerian civilization.

The Hindu Vedas, a collection of ancient Indian texts, contain numerous references to celestial beings and advanced technology that some proponents of Ancient Alien Theory argue could be interpreted as evidence of extraterrestrial visitation. For instance, the Rigveda describes flying machines known as vimanas, which were used by the gods to traverse the skies and even engage in aerial battles. These descriptions of advanced aerial technology, dating back thousands of years, have led some researchers to suggest that ancient Indians may have witnessed extraterrestrial technology in action.

The Hebrew Bible, also known as the Old Testament, contains several accounts that proponents of Ancient Alien Theory believe could be interpreted as encounters with extraterrestrial beings. The story of Ezekiel's vision, for example, describes a flying chariot with wheels and shining metal, accompanied by celestial beings. Some researchers argue that this account could be an ancient description of an extraterrestrial spacecraft and its occupants. Additionally, the story of the Nephilim, a race of giants mentioned in the Book of Genesis, has been suggested by some to be the result of extraterrestrial beings interbreeding with humans.

Various Native American legends also contain stories of celestial beings and advanced technology that proponents of Ancient Alien Theory argue could be evidence of extraterrestrial visitation. For example, the Hopi tribe of the southwestern United States tells of the Kachinas, spiritual beings who descended from the sky to teach the Hopi people various aspects of their culture, including agriculture and ceremonial practices. Similarly, the Dogon tribe of West Africa has a complex cosmology that includes detailed knowledge of the Sirius star system, which they claim was given to them by celestial beings known as the Nommo. These accounts, which predate modern astronomical knowledge, have led some researchers to suggest that ancient Native American tribes may have had direct contact with extraterrestrial beings.

In addition to religious and mythological texts, ancient historical records, such as the annals of the Chinese, Egyptian, and Mesopotamian civilizations, contain accounts of unexplained aerial phenomena and encounters with otherworldly beings. These accounts often describe advanced technology and capabilities beyond the understanding of the time, leading some researchers to suggest that they could be evidence of ancient alien contact.

For example, ancient Chinese texts contain numerous references to flying objects and celestial beings. The "Bamboo Annals," a historical record of the Xia and Shang dynasties, mentions instances of "dragons" appearing in the sky, which some researchers argue could be a description of extraterrestrial spacecraft. Similarly, ancient Egyptian texts, such as the "Tulli Papyrus," describe sightings of fiery disks in the sky, which some proponents of Ancient Alien Theory interpret as evidence of extraterrestrial visitation.

The ancient civilizations of Mesopotamia, including the Sumerians, Babylonians, and Assyrians, left behind records that contain accounts of otherworldly beings and unexplained phenomena. The Enuma Elish, a Babylonian creation myth, speaks of a race of gods who came from the heavens and created humans to serve them. Some proponents of Ancient Alien Theory argue that these gods may represent extraterrestrial beings who influenced the development of ancient Mesopotamian cultures. Additionally, the Epic of Etana, an ancient Sumerian tale, describes a king who is carried into the heavens by an eagle, possibly representing an encounter with advanced extraterrestrial technology.

The analysis of ancient texts and mythology from diverse cultures provides a wealth of accounts that describe encounters with celestial beings, advanced technology, and unexplained phenomena. Proponents of Ancient Alien Theory argue that the similarities between these accounts, as well as the advanced knowledge and capabilities they describe, provide compelling evidence of extraterrestrial visitation and influence on ancient human civilizations. As our understanding of human history and the universe continues to evolve, the exploration of ancient texts and mythology will undoubtedly remain a crucial element in the ongoing debate surrounding Ancient Alien Theory.

The concept of ancient aliens has not only permeated popular culture and science fiction but has also played a significant role in shaping public interest in the subject. By presenting compelling narratives and exploring the idea of extraterrestrial beings visiting and influencing human history, various forms of media have contributed to the popularization of Ancient Alien Theory and its continued presence in the public consciousness.

Films such as "Prometheus" (2012), directed by Ridley Scott, have captured the imaginations of audiences by presenting the idea that ancient aliens may have played a role in the creation of human life. The film's storyline revolves around a team of explorers who discover a star map in ancient Earth artifacts, leading them to a distant planet where they encounter an advanced extraterrestrial race. The themes explored in the movie provoke thought and discussion regarding the potential existence of ancient aliens and their influence on human history.

The long-running television series "Ancient Aliens" has been instrumental in popularizing the theory, as it presents a wide range of evidence and expert opinions on the subject. This documentary series delves into various archaeological sites, artifacts, and historical accounts that proponents of the theory believe provide evidence of extraterrestrial visitation. By presenting these findings in an accessible and engaging format, "Ancient Aliens" has introduced the concept to a broad audience and stimulated further interest and research into the topic.

Science fiction works such as H.P. Lovecraft's Cthulhu Mythos, a series of interconnected stories centered around ancient extraterrestrial beings known as the "Great Old Ones," have further fueled public interest in Ancient Alien Theory. Lovecraft's tales, which describe the Great Old Ones as powerful entities that once ruled the Earth and continue to exert their influence on humanity, provide a fascinating and imaginative exploration of the idea that ancient aliens may have played a role in shaping human history.

The "Stargate" franchise, which includes the 1994 film "Stargate" and its subsequent television series "Stargate SG-1," "Stargate Atlantis," and "Stargate Universe," has also contributed to the popularization of Ancient Alien Theory. The franchise's central premise involves the discovery of ancient alien technology in the form of stargates, which enable instantaneous travel between distant planets. Throughout the series, human explorers encounter various advanced extraterrestrial races and civilizations, many of which have had a profound impact on human history and culture. The "Stargate" franchise thus presents an intriguing and entertaining exploration of the potential interactions between ancient aliens and human civilizations.

Popular novels, such as Chariots of the Gods? by Erich von Däniken and The 12th Planet by Zecharia Sitchin, have also played a crucial role in shaping public interest in Ancient Alien Theory. These works, which present compelling arguments and interpretations of archaeological and historical evidence in favor of the existence of ancient aliens, have inspired countless readers to question conventional explanations of human history and consider the possibility of extraterrestrial intervention.

In conclusion, the impact of popular culture and science fiction on Ancient Alien Theory has been significant, as films, television series, books, and other media have introduced the concept to a wide audience and stimulated public interest in the subject. By presenting compelling narratives and exploring the potential existence of ancient aliens and their influence on human history, these works have contributed to the continued popularity and relevance of Ancient Alien Theory in the public consciousness.

Despite the growing interest in and popularity of Ancient Alien Theory, it has not been without its critics. Skeptics argue that the evidence supporting the existence of ancient aliens is often based on speculation, misinterpretation, or the dismissal of conventional explanations. Many experts in the fields of archaeology, history, and anthropology contend that the accomplishments of ancient civilizations can be explained without invoking extraterrestrial intervention.

However, proponents of Ancient Alien Theory maintain that the existing evidence, when viewed collectively, presents a compelling case for the existence of ancient extraterrestrial beings and their influence on human history. They argue that critics often dismiss or overlook crucial pieces of evidence, such as the advanced engineering of ancient structures or the detailed accounts of celestial beings found in ancient texts.

For instance, while critics may claim that the construction of the Great Pyramids of Giza can be explained through conventional means, proponents of Ancient Alien Theory point to the precision and scale of these structures, which defy easy explanation. They argue that the use of advanced technology, potentially provided by extraterrestrial beings, may have been necessary for the construction of such awe-inspiring monuments.

Another area of contention is the interpretation of ancient texts and mythologies. Critics often argue that the stories of celestial beings found in these texts are merely symbolic or allegorical in nature. However, proponents of Ancient Alien Theory assert that these accounts should be taken more literally, as they often contain descriptions of advanced technology and knowledge that seems beyond the capabilities of ancient civilizations.

Furthermore, proponents of Ancient Alien Theory emphasize that critics often rely on incomplete or outdated information when evaluating the theory. As new discoveries are made in the fields of archaeology, astronomy, and astrobiology, the case for ancient aliens becomes increasingly stronger. For example, the recent discovery of potentially habitable exoplanets and advancements in our understanding of extremophiles have lent further credibility to the idea that life could exist elsewhere in the universe and may have visited Earth in the past.

In addition, proponents of Ancient Alien Theory argue that the dismissal of unconventional explanations is a result of academic bias and a reluctance to challenge established paradigms. They contend that an open-minded approach, which considers all available evidence, is necessary to advance our understanding of human history and the potential existence of extraterrestrial life.

While Ancient Alien Theory has faced significant criticisms and controversies, proponents maintain that the collective body of evidence presents a compelling case for the existence of ancient extraterrestrial beings and their influence on human history. As scientific advancements continue to shed light on the mysteries of our past and the vast expanse of the cosmos, the debate surrounding Ancient Alien Theory is likely to persist, encouraging further research and exploration into this fascinating subject.

The origins and history of Ancient Alien Theory stretch far back through the annals of human history, encompassing a wealth of cultural perspectives, disciplines, and ideas. From the ancient mythologies and religious texts that speak of otherworldly beings and their interactions with humanity, to the modern fields of archaeology, astronomy, and astrobiology, which reveal an ever-expanding universe teeming with possibilities, Ancient Alien Theory remains an intriguing and controversial subject.

While critics argue that the evidence supporting Ancient Alien Theory is speculative or based on misinterpretations, proponents assert that the collective body of evidence, when viewed through a lens that embraces the potential for extraterrestrial life, presents a compelling narrative of ancient alien involvement in human history. From the seemingly impossible feats of engineering found in ancient structures to the enigmatic artifacts that defy conventional explanations, the mysteries of our past appear to suggest the guiding hand of advanced civilizations from beyond our world.

As new discoveries are made both on Earth and throughout the cosmos, Ancient Alien Theory will continue to evolve and adapt, inspiring further debate, research, and exploration into this fascinating subject. With each new piece of evidence, we come closer to understanding our place in the vast tapestry of the universe and the potential for intelligent life beyond our planet.

The advancements in technology and our understanding of the universe are rapidly expanding, bringing with them the potential for definitive proof of the existence of extraterrestrial life. If discovered, such proof would lend immense credibility to the idea that ancient aliens played a role in shaping our past and molding the trajectory of human civilization. Until then, Ancient Alien Theory will continue to captivate the imaginations of countless individuals, encouraging us to question the conventional understanding of our history and our place in the cosmos.

In conclusion, the origins and history of Ancient Alien Theory offer a fascinating window into humanity's enduring fascination with the possibility of extraterrestrial life and its potential impact on our world. By examining the evidence, stories, and theories presented throughout history, we can appreciate the significance of this concept and its potential to reshape our perception of the past, present, and future. As we continue to explore the mysteries of our universe and our own history, Ancient Alien Theory will remain an intriguing and thought-provoking subject, challenging our understanding of what we know and what we have yet to discover.

With each step forward, we inch closer to unlocking the secrets of the cosmos and uncovering the truth about the potential role of ancient aliens in our world. The study of Ancient Alien Theory, which blends history, mythology, and science, serves as a reminder of the limitless potential for discovery that awaits us, inviting us to ponder the mysteries of the universe and embrace the possibility that we are not alone in the vast expanse of space.

Chapter 2

Theories of the Ancient Astronaut Hypothesis

the Concepts and Principles

The Ancient Astronaut Hypothesis proposes that Earth has been visited by extraterrestrial beings in the past, who played a significant role in the development of human civilizations, imparting knowledge, technology, and cultural influence. This hypothesis is based on various pieces of evidence, including archaeological findings, ancient texts, and similarities between myths and legends from different cultures. This chapter aims to explore the various theories and aspects of the Ancient Astronaut Hypothesis, providing a comprehensive understanding of its implications and significance.

The Ancient Astronaut Hypothesis is based on the theoretical foundation that not only does life exist beyond Earth, but intelligent extraterrestrial beings have the ability and motivation to visit our planet. Three key factors lend credence to this idea: the vastness of the universe, the discovery of potentially habitable exoplanets, and the ongoing search for extraterrestrial life through projects like SETI (Search for Extraterrestrial Intelligence).

The sheer size and scale of the universe make it statistically probable that there are other planets hosting life. There are an estimated 100 billion galaxies in the observable universe, each containing hundreds of billions of stars. With such a staggering number of stars, it is reasonable to assume that many of them have planets orbiting them. The vastness of the universe significantly increases the likelihood that there are other forms of intelligent life out there, some of which may possess the capability to visit Earth.

The ongoing discovery of exoplanets—planets outside our solar system—has revealed numerous candidates that may possess the conditions necessary to support life. Some of these planets exist within their star's habitable zone, the region around a star where conditions are just right for liquid water to exist on a planet's surface. Water is considered a crucial ingredient for life as we know it, and the presence of liquid water on a planet increases the likelihood of life developing there.

As of September 2021, over 4,000 exoplanets have been confirmed, with more being discovered on a regular basis. The Kepler Space Telescope, launched in 2009, has been particularly instrumental in identifying potentially habitable exoplanets. The growing list of Earth-like planets bolsters the idea that intelligent extraterrestrial life could exist elsewhere in the universe, potentially possessing advanced technology that enables them to visit our planet.

Projects like SETI (Search for Extraterrestrial Intelligence) actively search for evidence of intelligent life beyond Earth. SETI primarily focuses on detecting radio signals from extraterrestrial civilizations, which could indicate the presence of technologically advanced beings. Although no definitive proof of extraterrestrial intelligence has been found, the ongoing efforts of SETI and other similar projects demonstrate the scientific community's belief in the possibility of intelligent life beyond our planet.

Moreover, recent developments in the search for extraterrestrial life include the study of extremophiles, organisms that thrive in extreme environments on Earth. These organisms have expanded our understanding of the possible conditions under which life can exist, suggesting that extraterrestrial life might be more diverse and widespread than previously assumed.

The Ancient Astronaut Hypothesis is deeply rooted in the theoretical foundation that intelligent extraterrestrial life exists and has the capacity to visit Earth. The vastness of the universe, the discovery of potentially habitable exoplanets, and the ongoing search for extraterrestrial life through projects like SETI provide compelling reasons to consider the possibility that ancient aliens have influenced human history and culture. As our understanding of the cosmos and the potential for extraterrestrial life continues to grow, the Ancient Astronaut Hypothesis remains a captivating area of study that challenges conventional explanations of human history and development.

The concept of paleovisits and paleocontact is central to the Ancient Astronaut Hypothesis, as it suggests that extraterrestrial beings played a significant role in shaping human history. Advocates of this hypothesis argue that ancient aliens not only visited Earth but also interacted with early human civilizations, leading to significant advancements in culture, technology, and biology. This section aims to expand on the idea of paleovisits and paleocontact, exploring the various ways in which ancient aliens may have influenced humanity.

Proponents of the Ancient Astronaut Hypothesis argue that the visits of extraterrestrial beings led to rapid cultural advancements in various ancient civilizations. They suggest that the sudden emergence of complex societies, sophisticated systems of writing, and advanced knowledge in fields such as astronomy, mathematics, and medicine can be attributed to the direct or indirect influence of ancient aliens.

For instance, the Sumerian civilization, one of the oldest known human societies, is said to have developed a complex system of writing, advanced agricultural techniques, and impressive architectural structures within a relatively short period. Some researchers argue that this rapid development is evidence of extraterrestrial intervention, as the Sumerians themselves attributed their advancements to the knowledge imparted by the Anunnaki, a race of gods who descended from the heavens.

Ancient astronaut theorists also argue that paleovisits led to significant technological advancements in various ancient civilizations. Examples of such advancements include the construction of megalithic structures, the development of advanced metallurgy, and the creation of complex machines and devices that were seemingly ahead of their time.

One notable example is the Antikythera mechanism, an ancient Greek device believed to be an astronomical calculator that dates back to the 2nd century BCE. The complexity and precision of this device are astonishing, considering the technology available at the time. Proponents of the Ancient Astronaut Hypothesis argue that such advanced technology could only have been developed with the help of extraterrestrial beings.

The idea of ancient aliens influencing the biological development of humanity is another aspect of paleocontact. According to this theory, extraterrestrial beings may have manipulated the genetic makeup of early humans, either to create a hybrid species or to enhance human capabilities.

This theory is supported by various ancient myths and legends that describe gods mating with humans or creating humans through divine intervention. Some proponents of the Ancient Astronaut Hypothesis also argue that the sudden emergence of Homo sapiens and their rapid cognitive development can be attributed to extraterrestrial intervention, as it is difficult to explain this phenomenon through the conventional theory of evolution alone.

Proponents of the Ancient Astronaut Hypothesis point to various pieces of evidence to support the idea of paleovisits and paleocontact. This evidence includes ancient texts that describe encounters with divine beings, artifacts that depict advanced technology or extraterrestrial beings, and the presence of megalithic structures that defy conventional explanations.

One example is the Vimanas described in ancient Indian texts such as the Ramayana and the Mahabharata. These texts provide detailed accounts of flying chariots or vehicles, which some researchers argue are evidence of advanced technology or extraterrestrial spacecraft.

The concept of paleovisits and paleocontact is central to the Ancient Astronaut Hypothesis and provides a compelling alternative perspective on human history. If true, this theory would have profound implications for our understanding of human civilization, the origins of culture and technology, and the potential existence of extraterrestrial life.

The Direct Intervention theory is a significant aspect of the Ancient Astronaut Hypothesis, which proposes that extraterrestrial beings played an active role in shaping human history by providing advanced knowledge, technology, and guidance to early civilizations. This section will delve deeper into the examples of direct intervention, focusing on megalithic structures and various instances where ancient alien involvement seems highly plausible.

One of the most compelling arguments in favor of the Direct Intervention theory is the existence of megalithic structures around the world. These ancient monuments, such as the Pyramids of Giza, the statues on Easter Island (Moai), Puma Punku, and Göbekli Tepe, exhibit extraordinary architectural and engineering feats that are difficult to explain using conventional historical knowledge.

For instance, the Great Pyramid of Giza was built with such precision that the four sides of the pyramid are aligned to the cardinal points within a fraction of a degree. Furthermore, the massive stone blocks used in its construction, some weighing up to 70 tons, were quarried from a location over 500 miles away. The statues on Easter Island, or Moai, also represent an extraordinary engineering achievement. These massive stone figures, weighing up to 75 tons each, were transported over considerable distances without the use of wheels, and were erected on large stone platforms called ahu.

The complexity and precision exhibited in these structures suggest that advanced technology, beyond the capabilities of ancient humans, may have been involved in their construction. Proponents of the Direct Intervention theory argue that extraterrestrial beings provided the necessary knowledge and tools to accomplish these feats, which would have been impossible to achieve otherwise.

The Direct Intervention theory also postulates that ancient astronauts imparted advanced knowledge to early human civilizations, enabling them to develop complex technologies and make remarkable scientific discoveries. This transfer of knowledge is evident in various ancient texts, which contain surprisingly accurate descriptions of the solar system, mathematics, and advanced engineering principles.

For example, the Sumerian civilization, which existed over 6,000 years ago, was remarkably advanced in astronomy. Their knowledge of celestial bodies, their movements, and their impact on Earth's climate and agriculture is astonishing, considering the limited tools available at that time. Similarly, the ancient Indian texts, such as the Vedas and the Mahabharata, contain detailed descriptions of advanced technologies, including flying machines called Vimanas and powerful weapons that resemble modern nuclear devices.

These examples of advanced knowledge transfer indicate that early human civilizations may have been in contact with extraterrestrial beings, who provided them with the necessary information to develop sophisticated technologies and make significant scientific discoveries.

In addition to providing advanced knowledge, the Direct Intervention theory suggests that ancient astronauts may have collaborated with early human civilizations to develop complex technologies and construct monumental structures. This collaboration could have taken the form of technological assistance, such as providing advanced tools or machinery, or direct involvement in the construction process.

One notable example is the Baalbek temple complex in Lebanon, which features the Trilithon, a set of three colossal stone blocks weighing between 800 and 1,200 tons each. These massive stones were quarried and transported several miles before being lifted and positioned with incredible precision. The logistics and engineering skills required for such an undertaking are difficult to fathom, even with modern technology.

Similar examples of extraordinary engineering can be found in other ancient sites, such as Sacsayhuamán in Peru, where massive stone blocks weighing up to 200 tons were cut and fitted together with laser-like precision, creating a fortress that has stood for over a thousand years.

The Cultural Influence theory suggests that ancient astronauts played a significant role in shaping human societies by indirectly inspiring religious beliefs, myths, and cultural practices. Proponents of this theory argue that the prevalence of similar myths and legends across various ancient civilizations is a strong indication of extraterrestrial contact in the distant past. This section will explore the key aspects of the Cultural Influence theory in greater detail.

The foundation of the Cultural Influence theory lies in the striking similarities found in the mythologies and religious beliefs of ancient civilizations across the globe. These commonalities often involve gods or divine beings descending from the heavens, interacting with humans, and bestowing knowledge, wisdom, or advanced technology upon them.

Examples of such parallels include the gods of ancient Greece, who resided on Mount Olympus and frequently intervened in human affairs, and the Egyptian gods, who were believed to have descended from the sky and played a crucial role in the development of Egyptian civilization. Similarly, the deities of various Mesoamerican cultures, such as the Maya and the Aztecs, were often associated with the heavens and were believed to possess advanced knowledge and capabilities.

Proponents of the Cultural Influence theory suggest that these similarities may not be mere coincidences but rather evidence of actual encounters with extraterrestrial beings. They argue that the myths and legends from these diverse cultures may represent distorted accounts of real events, in which ancient astronauts visited Earth and interacted with early human societies.

Another key aspect of the Cultural Influence theory is the presence of flying vehicles or chariots in numerous ancient myths and legends. These accounts describe gods or divine beings traveling through the skies in fantastic vehicles, often accompanied by bright lights, sounds, or other extraordinary phenomena.

For example, the Indian epic Ramayana describes flying chariots called Vimanas, which were used by the gods to traverse the heavens and wage battles. Similarly, the ancient Chinese myth of the Yellow Emperor, Huangdi, tells of a divine being who arrived on Earth in a "dragon-boat" capable of flight. The Biblical account of the prophet Ezekiel also describes a vision of a heavenly chariot, which some ancient astronaut theorists interpret as an encounter with an extraterrestrial spacecraft.

Advocates of the Cultural Influence theory argue that these stories may be early descriptions of advanced technology, such as aircraft or spacecraft, provided by extraterrestrial visitors. They contend that the remarkable similarities in these accounts from diverse cultures suggest a common origin, pointing to encounters with ancient astronauts as a plausible explanation.

In addition to myths and legends, various ancient artworks and artifacts appear to depict extraterrestrial beings or advanced technology. These depictions are often found in the form of cave paintings, sculptures, or architectural reliefs and can be seen across different cultures and time periods.

Examples include the ancient Aboriginal rock art of Australia, which portrays humanoid figures with elongated heads and large, almond-shaped eyes, similar to the stereotypical depiction of extraterrestrial beings known as "greys." Additionally, the ancient Mayan city of Palenque features a carving commonly referred to as the "Palenque Astronaut," which appears to depict a figure seated inside a complex, technologically advanced vehicle.

Proponents of the Cultural Influence theory argue that these artistic depictions provide further evidence of extraterrestrial contact in the past. They contend that these images may represent early human attempts to portray encounters with ancient astronauts, who visited Earth and influenced the development of human societies.

The Cultural Influence theory offers a compelling perspective on the potential impact of ancient astronauts on early human societies. By examining the similarities between myths, legends, and artistic depictions across diverse cultures, proponents of this theory suggest that extraterrestrial beings may have played a significant role in shaping our beliefs, traditions, and cultural practices. While definitive evidence remains elusive, the Cultural Influence theory provides a fascinating lens through which to view our shared human history and explore the possibility of ancient encounters with otherworldly beings.

The Cultural Influence theory is further supported by the presence of sacred sites and ancient structures with precise astronomical alignments. These sites, found all around the world, often exhibit advanced knowledge of astronomy and the cosmos, which proponents of the theory argue could have been provided by extraterrestrial visitors.

Examples of such sites include the Great Pyramids of Giza, which are aligned with the stars of Orion's Belt, and the ancient city of Teotihuacan in Mexico, whose layout appears to represent the positions of celestial bodies in the solar system. Additionally, Stonehenge in England and the Nazca Lines in Peru both exhibit precise alignments with the solstices and other astronomical events.

Advocates of the Cultural Influence theory argue that the advanced astronomical knowledge displayed at these sites is indicative of extraterrestrial influence. They suggest that ancient astronauts may have shared their knowledge of the cosmos with early human societies, inspiring the construction of these sacred sites and structures as a means of honoring or commemorating their celestial visitors.

The potential influence of ancient astronauts on early human societies has far-reaching implications for our understanding of human history, as well as our beliefs, values, and cultural practices. By considering the possibility that our ancestors interacted with extraterrestrial beings, we gain a new perspective on the origins of our myths, legends, and religious beliefs, and a deeper appreciation for the interconnectedness of human societies throughout history.

Furthermore, the Cultural Influence theory encourages us to consider the potential implications of future contact with extraterrestrial civilizations. If ancient astronauts did visit Earth and shape the development of our cultures, it raises important questions about the nature of our relationship with other intelligent beings in the universe and the potential impact of such encounters on our society and global consciousness.

The Cultural Influence theory offers a captivating perspective on the potential role of ancient astronauts in shaping human history and culture. While the evidence for this theory remains a subject of debate and further study, it provides a thought-provoking lens through which to explore the mysteries of our past and consider the possibilities for our future interactions with intelligent life beyond Earth.

The Genetic Manipulation theory is an intriguing aspect of the Ancient Astronaut Hypothesis, as it offers an explanation for the rapid development of Homo sapiens and the prevalence of certain myths and legends across various cultures. This theory suggests that ancient astronauts may have directly intervened in the genetic makeup of early humans, either to create a hybrid species or to enhance human capabilities.

A significant amount of evidence supporting the Genetic Manipulation theory comes from ancient myths and legends that describe gods or divine beings mating with humans or creating humans through divine intervention. In Greek mythology, for example, Zeus is known to have fathered numerous demigods through relationships with mortal women. Similarly, in Norse mythology, gods such as Odin and Thor are said to have sired human offspring.

These stories of divine-human relationships are found in cultures all around the world, suggesting a common theme of interaction between celestial beings and humans. Proponents of the Genetic Manipulation theory argue that these myths may be symbolic representations of actual genetic manipulation events carried out by extraterrestrial beings in the distant past.

The sudden emergence of Homo sapiens as a distinct species and their rapid cognitive development is another key piece of evidence supporting the Genetic Manipulation theory. Anthropological and archaeological research has shown that modern humans appeared relatively suddenly in the fossil record, approximately 300,000 years ago, with a significant leap in cognitive abilities compared to other hominid species.

Ancient Astronaut theorists propose that this rapid development could be the result of extraterrestrial intervention in human genetics. By manipulating the genetic makeup of early humans, ancient astronauts may have accelerated the evolution of human intelligence and consciousness, enabling the rapid development of advanced civilizations.

The existence of advanced capabilities in ancient civilizations, such as sophisticated astronomical knowledge, complex mathematical systems, and remarkable feats of engineering, may also be attributed to the Genetic Manipulation theory. If ancient astronauts enhanced human cognitive abilities through genetic manipulation, it could explain the extraordinary achievements of ancient civilizations that seem to surpass the capabilities of their contemporaries.

While direct genetic evidence for extraterrestrial intervention in human evolution is yet to be found, ongoing advancements in genetic research may one day uncover signs of manipulation in our DNA. Some researchers argue that specific genetic traits, such as the FOXP2 gene, which is responsible for language development, could be the result of ancient astronaut intervention.

The Genetic Manipulation theory offers an alternative explanation for the rapid development of Homo sapiens and the prevalence of divine-human interaction myths in various cultures. While conclusive evidence is still lacking, the possibility that ancient astronauts may have played a role in shaping human genetics remains a fascinating area of study within the broader Ancient Astronaut Hypothesis.

The Watchers theory postulates that ancient astronauts were primarily observers or "watchers" who monitored human progress without directly intervening in human affairs. This theory suggests that extraterrestrial visitors may have been driven by curiosity or had specific reasons to observe humanity's development from a distance. The Watchers theory presents a fascinating perspective on the potential interactions between ancient aliens and early human civilizations and provides intriguing insights into the possible motives of these extraterrestrial beings.

One possible explanation for the Watchers' non-interventionist approach is that they adhered to a policy or principle that prevented them from directly influencing human affairs. This policy could have stemmed from an ethical or philosophical belief in the importance of allowing civilizations to develop independently, without outside interference. This concept has parallels in modern science fiction, such as the "Prime Directive" in the Star Trek universe, which prohibits interfering with the natural development of other civilizations.

The idea of a non-interventionist policy suggests that ancient astronauts may have been a highly advanced and morally conscious civilization, keen to avoid the potential negative consequences of interfering with the development of other species. This notion adds depth to the Ancient Astronaut Hypothesis, implying that extraterrestrial beings were not only technologically advanced but also guided by ethical considerations.

Another possibility is that the Watchers were primarily driven by a scientific research agenda, studying human civilizations as part of a broader examination of life in the universe. In this context, the ancient astronauts may have been akin to anthropologists or sociologists, observing and documenting the development of human societies, technology, and culture from a distance.

If this were the case, the ancient astronauts might have been interested in understanding the diverse range of evolutionary paths, social structures, and technological advancements exhibited by various civilizations throughout the cosmos. By observing humanity, the Watchers could have gained valuable insights into the factors that contribute to the development of intelligent life and the emergence of complex societies.

The Watchers theory is supported by numerous accounts in ancient texts and mythologies that describe divine or otherworldly beings observing human affairs from afar. For example, the biblical Book of Enoch details a group of divine beings called "The Watchers," who were tasked with observing and reporting on human activities. Similarly, many ancient mythologies feature gods or celestial beings who watch over humanity from the heavens.

Some researchers argue that these accounts may be based on actual encounters with extraterrestrial beings who were observing human societies from a distance. In this interpretation, the Watchers may have appeared as mysterious figures in the sky or as elusive beings who occasionally made their presence known, only to vanish without a trace.

The Watchers theory provides a compelling alternative to other Ancient Astronaut Hypothesis theories, suggesting that ancient aliens may have taken a more passive and observational role in their interactions with early human civilizations. This perspective raises intriguing questions about the motives and ethics of these extraterrestrial beings and adds depth to our understanding of the potential relationship between ancient astronauts and humanity.

The construction of megalithic structures around the world has long been a subject of fascination and intrigue. Many of these structures exhibit exceptional architectural and engineering prowess, leading proponents of the Ancient Astronaut Hypothesis to argue that their construction would have been impossible without the aid of advanced technology provided by extraterrestrial visitors.

The Great Pyramid of Giza, located in Egypt, is one of the most iconic and enigmatic megalithic structures in the world. It is the oldest and largest of the three pyramids in the Giza pyramid complex and was built around 2560 BCE. The precision with which the pyramid was constructed is remarkable, with each of its four sides measuring almost equal lengths, and the angle of its slope being consistent throughout. The structure's alignment to true north is another extraordinary feature, with an accuracy of just 3/60th of a degree off.

The transport and positioning of the massive stone blocks used to build the Great Pyramid, some weighing up to 70 tons, is another aspect that challenges conventional explanations. The level of precision and coordination required to quarry, transport, and place these massive stones in perfect alignment would have been an extraordinary feat, even by today's standards. Proponents of the Ancient Astronaut Hypothesis argue that the construction of the Great Pyramid would have required advanced knowledge and technology, potentially provided by extraterrestrial visitors.

Stonehenge, located in England, is another enigmatic megalithic structure that has captured the imagination of researchers and enthusiasts for centuries. It is believed to have been built in several stages between 3000 and 2000 BCE. The monument consists of a circular arrangement of massive stones, some weighing up to 30 tons, and transported from distances up to 180 miles away.

The precise purpose of Stonehenge remains a mystery, but its astronomical alignments suggest that it may have functioned as a celestial observatory or calendar. The monument is aligned with the summer solstice sunrise and the winter solstice sunset, indicating an advanced knowledge of astronomy. Advocates of the Ancient Astronaut Hypothesis argue that the construction and precise alignments of Stonehenge could be the result of guidance and assistance from extraterrestrial beings.

The Nazca Lines, located in the Nazca Desert in southern Peru, are a series of massive geoglyphs created by removing the reddish-brown iron oxide-coated pebbles that cover the surface of the Nazca desert and revealing the light-colored earth underneath. These geoglyphs depict various animals, plants, and geometric shapes and are believed to have been created between 500 BCE and 500 CE.

The true purpose of the Nazca Lines remains a mystery, but their scale and complexity have led some proponents of the Ancient Astronaut Hypothesis to suggest that they may have been created as a form of communication or landing markers for extraterrestrial visitors. The fact that the full extent of the Nazca Lines can only be appreciated from the air adds to the intrigue surrounding their origin and has fueled speculation that they were intended to be viewed by beings capable of flight.

The remarkable engineering and architectural feats displayed in the construction of these megalithic structures provide compelling evidence in favor of the Ancient Astronaut Hypothesis. The precision, scale, and astronomical alignments of these structures continue to defy conventional explanations and suggest that advanced knowledge and technology, potentially provided by extraterrestrial visitors, may have played a role in their creation.

The discovery of various ancient artifacts that seem to depict advanced technology or extraterrestrial beings has fueled the debate surrounding the Ancient Astronaut Hypothesis. These artifacts, often found in disparate regions of the world, challenge conventional explanations and suggest that ancient civilizations may have had contact with advanced extraterrestrial beings. The following examples provide a closer look at some of these intriguing artifacts:

The Antikythera mechanism, discovered in a shipwreck off the coast of Greece, is an ancient Greek device believed to be an astronomical calculator. This complex and intricate artifact, dating back to the 2nd century BCE, is considered the world's first known analog computer. It has been suggested that the Antikythera mechanism was designed to predict astronomical positions and eclipses for calendrical and astrological purposes. The level of precision and sophistication exhibited by this device has led some proponents of the Ancient Astronaut Hypothesis to argue that it could only have been created with the help of advanced extraterrestrial knowledge.

The Baghdad Battery, discovered in modern-day Iraq, is a group of artifacts dated to the Parthian period, around 250 BCE to 224 CE. These artifacts consist of a ceramic pot, a copper cylinder, and an iron rod, which some researchers argue could have functioned as an ancient galvanic cell, producing a small electric current. While there is still debate about the purpose of the Baghdad Battery, its potential use as an electrical device demonstrates a level of technological understanding that was seemingly ahead of its time. Advocates of the Ancient Astronaut Hypothesis argue that this advanced knowledge may have been imparted by extraterrestrial visitors.

Numerous artifacts and artworks from ancient cultures around the world seem to portray flying vehicles, astronauts, or beings wearing what appears to be advanced technology.

Described in ancient Indian texts, such as the Vedas, Vimanas were flying chariots or vehicles used by the gods. These descriptions include intricate details of their propulsion systems, weaponry, and other advanced features that bear striking similarities to modern aircraft and spacecraft.

Found in Egypt, the Saqqara Bird is a wooden artifact dating back to around 200 BCE. It has a bird-like shape with wings, a tail, and a head that resembles the shape of an airplane. Some proponents of the Ancient Astronaut Hypothesis argue that this artifact represents a model of an ancient flying machine, suggesting that the Egyptians may have had knowledge of aeronautical principles long before the invention of the airplane.

Numerous ancient carvings and paintings depict humanoid figures wearing what appear to be helmets, suits, and other equipment reminiscent of modern astronauts or pilots. These artifacts have been found in various locations, such as the Mayan site of Palenque in Mexico and the caves of Tassili n'Ajjer in Algeria. Advocates of the Ancient Astronaut Hypothesis argue that these depictions provide evidence of extraterrestrial contact, as they seemingly portray beings with advanced technology or gear.

The existence of these ancient artifacts, which seemingly depict advanced technology or extraterrestrial beings, lends support to the Ancient Astronaut Hypothesis. These discoveries suggest that ancient civilizations may have had access to advanced knowledge or technology that could only have been provided by extraterrestrial visitors, challenging conventional historical narratives and opening up new possibilities for our understanding of human history.

Ancient texts from various civilizations across the globe provide intriguing descriptions of advanced technology, flying vehicles, and encounters with divine or otherworldly beings. Proponents of the Ancient Astronaut Hypothesis argue that these texts offer compelling evidence of extraterrestrial contact and influence in ancient human history.

The Indian epics, Ramayana and Mahabharata, contain numerous accounts of flying chariots called Vimanas, which were used by gods and heroes in epic battles and celestial journeys. These Vimanas were described as being powered by advanced propulsion systems, capable of incredible speeds, and even equipped with sophisticated weaponry. The detailed descriptions of Vimanas, their functioning, and capabilities found in these ancient texts suggest a level of technological understanding that far surpasses what would be expected from ancient civilizations.

Some proponents of the Ancient Astronaut Hypothesis argue that the Vimanas were not mere mythological constructs but rather representations of actual flying vehicles used by extraterrestrial beings who visited Earth during ancient times. They contend that these beings were possibly misinterpreted as gods by ancient humans, who subsequently incorporated their encounters with these advanced technologies into their religious and cultural narratives.

The Sumerian texts, dating back to one of the earliest known civilizations, provide intriguing accounts of a race of gods called the

Annunaki. These gods were said to have descended from the heavens to Earth and played a crucial role in the development of human civilization. The Annunaki were described as having taught humans agriculture, science, and other advanced knowledge.

The detailed descriptions of the Annunaki and their interactions with humans have led some proponents of the Ancient Astronaut Hypothesis to argue that these gods were, in fact, extraterrestrial beings who visited Earth during ancient times. The advanced knowledge attributed to the Annunaki may represent the transfer of technology and information from these extraterrestrial visitors to early human civilizations.

The Book of Enoch, an ancient Jewish text, contains accounts of a group of angelic beings called the Watchers, who descended to Earth and interacted with humans. These Watchers were said to have imparted advanced knowledge to humans, including metalworking, astrology, and other forms of esoteric wisdom. The text also describes the Watchers mating with human women, resulting in a race of hybrid beings called the Nephilim.

The accounts of the Watchers in the Book of Enoch bear striking similarities to the descriptions of the Annunaki in the Sumerian texts, which has led some proponents of the Ancient Astronaut Hypothesis to argue that both sets of accounts may represent encounters with the same extraterrestrial beings. The transfer of advanced knowledge and the creation of hybrid beings described in these texts could be interpreted as evidence of extraterrestrial intervention in human history.

Ancient texts from various civilizations contain intriguing accounts of advanced technology, flying vehicles, and otherworldly beings that defy conventional explanations. These texts provide a rich source of evidence for proponents of the Ancient Astronaut Hypothesis, who argue that these accounts represent encounters with extraterrestrial beings who visited Earth during ancient times, imparting their knowledge and influencing human development.

The Ancient Astronaut Hypothesis is also supported by cultural evidence, including the similarities between myths and legends from different civilizations across the globe. These stories often describe gods or divine beings descending from the heavens, interacting with humans, and providing them with knowledge and guidance.

The Anunnaki, as described in the Sumerian texts, were a race of divine beings who played a significant role in the development of human civilization. The Sumerians, one of the earliest known civilizations, are credited with many innovations, such as the development of the earliest known form of writing, known as cuneiform. The Sumerian civilization's advanced knowledge in various fields, coupled with their depictions of the Anunnaki, has led some proponents of the Ancient Astronaut Hypothesis to consider the possibility of an extraterrestrial origin for these enigmatic beings.

The Anunnaki were described as having human-like appearances but possessing extraordinary abilities and knowledge. They were believed to be immortal, and their superior wisdom and technology allowed them to rule over humanity. The Sumerian texts also describe the Anunnaki as having the capability to travel through space, suggesting that they may have come from a distant planet or star system.

The Anunnaki are often credited with providing humans with advanced knowledge in various fields, such as agriculture, science, architecture, and even spirituality. The development of these skills and technologies allowed human civilizations to flourish, and proponents of the Ancient Astronaut Hypothesis argue that this rapid progress would have been impossible without the guidance of extraterrestrial beings like the Anunnaki.

Examples of advancements attributed to the Anunnaki include the construction of the ziggurats, massive step-pyramid structures found throughout ancient Mesopotamia, which are believed to have served as temples and centers of learning. Additionally, the Sumerians developed advanced systems of mathematics, astronomy, and law, which some researchers argue may have been influenced by the Anunnaki.

Zecharia Sitchin, one of the leading proponents of the Ancient Astronaut Hypothesis, claimed that the Anunnaki originated from a distant planet called Nibiru. According to Sitchin's interpretation of the Sumerian texts, Nibiru is a part of our solar system but follows an elongated orbit, bringing it close to Earth only once every 3,600 years. Sitchin argued that the Anunnaki came to Earth in search of valuable resources, such as gold, which they needed to stabilize their planet's atmosphere.

While Sitchin's theories have been met with skepticism from mainstream academia, they have gained a large following among those who believe in the Ancient Astronaut Hypothesis. The idea of an advanced extraterrestrial civilization visiting Earth in the distant past and shaping human development continues to intrigue and captivate the imaginations of many.

The Anunnaki's potential extraterrestrial origins and their role in human history have influenced various aspects of modern culture. This includes inspiring the work of authors, filmmakers, and artists who explore themes related to ancient astronauts and their impact on humanity. The concept of the Anunnaki has also found its way into the works of various conspiracy theorists, who suggest that these beings continue to play a role in contemporary events and are part of a hidden agenda to control human civilization.

The Anunnaki, as described in the Sumerian texts, present a compelling case for the Ancient Astronaut Hypothesis. Their advanced knowledge, abilities, and potential extraterrestrial origins challenge conventional understanding of human history and raise important questions about our past and our place in the cosmos. While definitive proof remains elusive, the notion of ancient astronauts like the

Quetzalcoatl, the Mesoamerican deity revered by the Aztec, Toltec, and Maya civilizations, is often portrayed as a feathered serpent or dragon. This enigmatic figure is believed to have played a crucial role in shaping the development of these ancient cultures by imparting knowledge, wisdom, and technological advancements. The complex mythos surrounding Quetzalcoatl has led some researchers to propose that this deity may, in fact, be an ancient astronaut who visited Earth in the distant past.

Quetzalcoatl is often associated with the planet Venus, which was considered one of the most important celestial bodies in Mesoamerican astronomy. As a light-bringer or divine teacher, Quetzalcoatl is credited with introducing essential knowledge, such as agriculture, writing, the calendar, and the arts, to the people of Central America. This narrative bears striking similarities to other ancient astronaut theories, where extraterrestrial beings are said to have provided advanced knowledge to early human civilizations.

The legend of Quetzalcoatl also includes a prophecy of his eventual return to Earth. According to the myth, he departed the ancient city of Tula on a raft of snakes, promising to return one day. This aspect of the story has fueled speculation that Quetzalcoatl may have been an ancient astronaut who left Earth in a spacecraft, with the promise of a future visit.

The representation of Quetzalcoatl as a feathered serpent or dragon has led some researchers to draw parallels between this deity and other ancient astronaut theories, where extraterrestrial beings are often depicted as reptilian or serpentine in nature. Moreover, the feathered aspect of Quetzalcoatl could be interpreted as a reference to flight or advanced aerial technology, further supporting the ancient astronaut hypothesis.

Quetzalcoatl is also often depicted wearing elaborate headdresses and carrying strange objects, which could be interpreted as advanced technology or equipment used by extraterrestrial visitors. These depictions are reminiscent of other ancient cultures' portrayals of gods or divine beings wearing what appears to be advanced technology, further supporting the idea that Quetzalcoatl may have been an ancient astronaut.

The arrival of the Spanish conquistadors in Central America in the 16th century led to the first documented encounters between Europeans and the indigenous peoples of the region. The Spanish recorded detailed accounts of the native people's beliefs and traditions, including their reverence for Quetzalcoatl. Intriguingly, some of these accounts describe Quetzalcoatl as a light-skinned, bearded figure, which is unusual in Mesoamerican iconography and could suggest a non-indigenous origin for this deity.

The Spanish conquistador Hernán Cortés was even mistaken for the returning Quetzalcoatl by the Aztec emperor Moctezuma II, due to his light skin, facial hair, and advanced technology, including firearms and horses. This case of mistaken identity further demonstrates the possibility that the legend of Quetzalcoatl may have originated from encounters with an ancient astronaut, who was revered as a divine being due to his otherworldly appearance and advanced capabilities.

The legend of Quetzalcoatl, the feathered serpent deity of Mesoamerican cultures, provides a fascinating case study for the ancient astronaut hypothesis. The deity's association with advanced knowledge, celestial origins, and unusual physical characteristics lend credence to the idea that Quetzalcoatl may have been an extraterrestrial visitor who influenced the development of ancient Central American civilizations. The parallels between the Quetzalcoatl myth and other ancient astronaut theories from around the world suggest that these encounters with divine beings may represent a common thread in human history, where advanced extraterrestrial visitors imparted knowledge and technology to early human societies.

Several ancient Mesoamerican artifacts and architectural sites have been linked to the worship of Quetzalcoatl, some of which exhibit features that proponents of the ancient astronaut hypothesis argue could be evidence of extraterrestrial influence. Among these are the Temple of the Feathered Serpent at Teotihuacan, the Pyramid of Kukulcan at Chichen Itza, and various stone carvings and sculptures depicting Quetzalcoatl and other deities.

The Temple of the Feathered Serpent, for example, showcases intricate carvings of the feathered serpent deity, as well as mysterious, geometric patterns that some researchers argue could represent advanced mathematical or astronomical knowledge. The Pyramid of Kukulcan, also known as El Castillo, is another iconic structure associated with Quetzalcoatl, and its precise alignment with the sun during the spring and autumn equinoxes creates a visual illusion of a serpent slithering down its staircase. This remarkable feat of engineering has led some to speculate that the builders of these structures may have had access to advanced knowledge or technology provided by extraterrestrial visitors.

Another aspect that lends support to the ancient astronaut hypothesis is the existence of similar myths and legends in other cultures across the world. The figure of the feathered serpent is not unique to Mesoamerican cultures, with parallels found in other ancient civilizations, such as the Egyptian god Serapis, the Chinese dragon, and the biblical seraphim. These cross-cultural connections suggest that ancient human societies may have had common encounters with otherworldly beings, who were subsequently revered as divine figures.

Furthermore, the theme of a divine teacher or light-bringer who imparts advanced knowledge to human civilizations is prevalent in various mythologies, such as the Sumerian god Enki, the Greek god Prometheus, and the Hindu deity Vishnu. These similarities imply that the legend of Quetzalcoatl and the ancient astronaut hypothesis may be part of a larger pattern of extraterrestrial contact and influence throughout human history.

The figure of Quetzalcoatl, the feathered serpent deity of Mesoamerican cultures, provides a compelling case for the ancient astronaut hypothesis. The deity's association with advanced knowledge, celestial origins, and unique physical characteristics, as well as the intricate artifacts and architectural sites linked to his worship, all suggest that Quetzalcoatl may have been an extraterrestrial visitor who played a significant role in shaping the development of ancient Central American civilizations. By examining the legend of Quetzalcoatl and other similar myths from around the world, we can gain valuable insights into the potential existence of ancient alien contact and its impact on human history.

The Nephilim, as described in the Hebrew Bible, have long been a source of fascination and speculation. These enigmatic beings were said to be the offspring of "the sons of God" and human women, resulting in a race of giants with extraordinary abilities. Proponents of the Ancient Astronaut Hypothesis argue that the Nephilim may represent a hybrid species created through extraterrestrial genetic manipulation, a notion that is supported by several factors.

The Nephilim were described as possessing extraordinary strength and abilities that set them apart from ordinary humans. They were also considered to be much taller than the average person, a trait often associated with their status as giants. These unique characteristics have led some proponents of the Ancient Astronaut Hypothesis to argue that the Nephilim were the result of extraterrestrial beings breeding with human women, thereby imparting their own advanced capabilities and physical traits onto their hybrid offspring.

The concept of a race of giants born from the union of divine beings and humans is not unique to the Hebrew Bible. Similar stories can be found in various mythologies around the world, including Greek, Norse, and Hindu traditions. For example, the Greek mythology speaks of the demigods, who were the offspring of gods and humans. These demigods possessed extraordinary abilities and often played crucial roles in the affairs of gods and mortals.

These cross-cultural similarities lend credibility to the idea that the Nephilim may have been a global phenomenon, representing the result of extraterrestrial beings interbreeding with human populations across different geographical locations.

The Ancient Astronaut Hypothesis proposes that extraterrestrial beings visited Earth in the distant past and played an active role in shaping human civilization. The creation of the Nephilim as a hybrid species could be seen as part of this intervention, with the extraterrestrial beings seeking to enhance human capabilities by introducing their own genetic material into the human gene pool.

This idea is supported by various ancient texts and mythological accounts that describe gods or divine beings mating with humans or creating humans through divine intervention. The Sumerian texts, for example, describe the Annunaki as being responsible for creating humans to serve as their laborers, while the Greek myths recount numerous instances of gods mating with mortals and producing demigods as offspring.

There are several possible motivations for extraterrestrial beings to engage in genetic manipulation and create a hybrid species like the Nephilim. Some researchers suggest that this could have been an experiment to study the effects of combining different genetic traits or to create a superior species that could help advance the civilization of their human counterparts. Others argue that the creation of the Nephilim may have been an attempt by the extraterrestrial visitors to establish a ruling class or a group of intermediaries between themselves and the human populations they sought to influence.

The Nephilim present a fascinating and compelling case for the Ancient Astronaut Hypothesis. The unique characteristics, abilities, and cross-cultural similarities of these enigmatic beings suggest that they may have been the product of extraterrestrial genetic manipulation, providing further evidence for the idea that advanced alien beings played a significant role in shaping human history and civilization.

The Ancient Astronaut Hypothesis carries significant implications for our understanding of human history, the nature of the universe, and the potential existence of extraterrestrial life. By suggesting that ancient aliens played a crucial role in the development of human civilization, this hypothesis challenges conventional beliefs and encourages us to reevaluate our understanding of the origins of religion, mythology, and culture.

If ancient astronauts did, in fact, visit Earth and interact with early human civilizations, this could have profound implications for our understanding of religion and spirituality. Many religious texts and mythologies describe gods and divine beings descending from the heavens, providing guidance, knowledge, and even punishment. By interpreting these accounts as actual encounters with extraterrestrial beings, the Ancient Astronaut Hypothesis provides a new perspective on the origin and purpose of religious beliefs and practices, potentially reshaping our understanding of the divine.

The Ancient Astronaut Hypothesis also suggests that extraterrestrial beings may have introduced advanced technologies to early human civilizations, inspiring rapid scientific and technological progress. This notion could encourage researchers to investigate the possibility of rediscovering lost knowledge or technologies that may have been passed down from ancient astronauts. Furthermore, the hypothesis raises the question of whether current technological advancements, such as artificial intelligence, quantum computing, and advanced propulsion systems, could have been influenced by ancient extraterrestrial contact.

The possibility of future contact with extraterrestrial beings is another significant implication of the Ancient Astronaut Hypothesis. If ancient astronauts visited Earth in the past, it is plausible that they are still monitoring our progress and may choose to interact with us again in the future. This possibility has fueled the ongoing search for extraterrestrial intelligence (SETI) and has inspired a renewed interest in space exploration and the study of our cosmic origins.

The prospect of future contact with extraterrestrial beings raises important ethical, philosophical, and practical considerations. How should humanity prepare for such an encounter? What would be the consequences of establishing contact with a more advanced civilization? These questions have inspired various international initiatives, such as the United Nations' Committee on the Peaceful Uses of Outer Space (COPUOS), which aims to develop guidelines and protocols for potential encounters with extraterrestrial beings.

The Ancient Astronaut Hypothesis has already made a significant cultural and societal impact, inspiring countless works of science fiction, documentaries, and academic research. By offering an alternative perspective on human history, the hypothesis has captured the imagination of millions of people worldwide and has encouraged a broader interest in the exploration of our cosmic origins.

The Ancient Astronaut Hypothesis presents a compelling alternative perspective on human history, challenging conventional understanding and encouraging further exploration of the potential existence of extraterrestrial life. While many aspects of this hypothesis remain speculative and controversial, the ongoing investigation of ancient aliens and their potential impact on our past, present, and future continues to inspire researchers, enthusiasts, and the general public alike.

Ancient Astronaut Theorists are invaluable contributors to the ongoing discussion around the existence of ancient aliens and their potential impact on human history. Their work spans various disciplines, from archaeology and anthropology to history and mythology, and their expertise has helped to identify and interpret potential evidence for extraterrestrial contact in the past. These theorists, who are unafraid to challenge conventional academic thought, have brought new perspectives to the study of our past and the potential existence of extraterrestrial life.

Erich von Däniken is a Swiss author who is widely considered the father of the Ancient Astronaut Hypothesis, due to his groundbreaking book "Chariots of the Gods?" published in 1968. The book presents a range of evidence, from ancient artwork and artifacts to megalithic structures, suggesting that Earth was visited by extraterrestrial beings in the past. Despite facing skepticism from mainstream academia, von Däniken has continued to write and lecture on the subject, authoring over 40 books and inspiring a new generation of researchers to explore the Ancient Astronaut Hypothesis.

Zecharia Sitchin was an Azerbaijani-born American author and researcher who studied the Sumerian texts and developed the Anunnaki theory. According to Sitchin, the Anunnaki were a race of extraterrestrial beings who visited Earth approximately 450,000 years ago and were responsible for the creation of human civilization. Sitchin's work focused on the interpretation of ancient texts and the connections between ancient Sumerian mythology and the potential existence of advanced extraterrestrial life. Although his theories have been met with skepticism, Sitchin's work has inspired many researchers to delve deeper into the study of ancient texts and their potential extraterrestrial connections.

Giorgio A. Tsoukalos is a Greek-Swiss television personality, producer, and Ancient Astronaut Theorist. He is best known for his appearances on the History Channel's television series "Ancient Aliens," where he discusses various aspects of the Ancient Astronaut Hypothesis. Tsoukalos has also appeared in numerous documentaries and television programs exploring the subject and has worked as a consulting producer on various projects related to ancient astronaut theories. His charismatic and enthusiastic approach to the subject has helped to popularize the Ancient Astronaut Hypothesis and bring it to a wider audience.

In addition to these prominent figures, many other researchers and enthusiasts have contributed to the ongoing debate surrounding the Ancient Astronaut Hypothesis. Their work encompasses a wide range of topics, including the examination of ancient texts, the analysis of archaeological findings, and the study of cultural similarities between different civilizations.

These theorists have played a crucial role in promoting the Ancient Astronaut Hypothesis and challenging conventional explanations for various archaeological and cultural phenomena. Their dedication to exploring alternative perspectives has helped to uncover potential evidence for extraterrestrial contact in the past and has inspired a new generation of researchers to continue investigating this fascinating and controversial area of study. As our understanding of the universe and our own history continues to evolve, the work of these Ancient Astronaut Theorists will remain an invaluable resource for those seeking to uncover the potential role of extraterrestrial life in shaping human history.

The Ancient Astronaut Hypothesis has been met with skepticism and criticism from various quarters, including mainstream academia, historians, and archaeologists. Critics argue that the similarities between myths and legends from different civilizations can be explained through cultural diffusion, the process by which cultural elements are transmitted from one society to another. This explanation suggests that stories of gods and divine beings descending from the heavens may have originated in one culture and then spread to others through trade, conquest, or other forms of interaction.

Skeptics of the Ancient Astronaut Hypothesis maintain that the construction of megalithic structures and the development of advanced technology in the past can be explained through the ingenuity and resourcefulness of human beings, rather than the intervention of extraterrestrial beings. They argue that ancient civilizations were capable of remarkable feats of engineering and architecture using the tools and techniques available to them at the time.

Another common counterargument is that the Ancient Astronaut Hypothesis relies heavily on the interpretation of ancient texts, artifacts, and cultural phenomena, which are often open to multiple interpretations. Critics contend that proponents of the hypothesis may be engaging in selective interpretation or allowing their imaginations to influence their conclusions.

Despite these criticisms, the Ancient Astronaut Hypothesis remains a compelling and intriguing area of study for many researchers and enthusiasts. As our understanding of the universe and our own past continues to evolve, the hypothesis may yet offer valuable insights into the nature of human history and our place in the cosmos.

Chapter 3

The Extraterrestrial Hypothesis

Is There Evidence of Intelligent Life Beyond Earth?

The Extraterrestrial Hypothesis (ETH) proposes the existence of intelligent life beyond Earth and the possibility that such beings have visited our planet in the past. Advocates of the Ancient Alien Theory, which is an offshoot of ETH, argue that extraterrestrial beings played a significant role in the development of human civilization, sharing their knowledge and technology with our ancestors. In this chapter, we will explore the scientific evidence that supports the existence of intelligent extraterrestrial life and examine the potential connections to ancient aliens.

The Drake Equation and the Fermi Paradox serve as essential frameworks for the exploration of intelligent extraterrestrial life and the possibility of ancient aliens. The Drake Equation, in particular, provides a basis for estimating the number of intelligent civilizations that could have existed in the past or are currently active in our galaxy.

The Drake Equation: A Case for Ancient Aliens

By considering various factors, the Drake Equation offers an optimistic view of the potential for intelligent extraterrestrial life. Given the age of the universe (approximately 13.8 billion years), it is possible that advanced civilizations could have developed long before humans emerged on Earth. These civilizations may have possessed the technology and knowledge to visit our planet during the time of ancient human civilizations.

Even with the uncertainties associated with the input parameters, the Drake Equation still yields an estimate that suggests numerous advanced civilizations might have existed or currently exist in our galaxy. This provides a solid foundation for the idea that ancient aliens could have visited Earth, influencing the development of early human civilizations.

The Fermi Paradox raises an important question – if there are numerous intelligent civilizations, why have we not encountered them? This paradox has led to several proposed solutions that, when examined closely, provide further support for the Ancient Alien Theory.

One proposed solution to the Fermi Paradox is the Zoo Hypothesis, which suggests that extraterrestrial civilizations are aware of Earth and its inhabitants but are deliberately avoiding contact to allow humans to develop independently. If this is the case, it is possible that ancient aliens visited Earth in the past, providing guidance and knowledge to early human civilizations, but have since decided to limit their interactions with humanity.

Similar to the Zoo Hypothesis, the Prime Directive Hypothesis proposes that advanced civilizations avoid interfering with less developed civilizations, adhering to a code of non-interference. Ancient aliens may have visited Earth when human civilization was in its infancy, leaving behind traces of their presence in ancient texts, art, and structures. Since then, they may have chosen to observe our development from a distance, in line with the non-interference principle.

Another possible solution to the Fermi Paradox is that we have not encountered extraterrestrial civilizations due to our limited detection methods or because they are using communication technologies beyond our current understanding. Ancient aliens may have had the means to visit Earth in the past, but their methods of communication or travel could be vastly different from ours, making it difficult for us to detect their presence today.

The Drake Equation and the Fermi Paradox, when examined from the perspective of Ancient Alien Theory, provide a compelling case for the existence of ancient aliens. The estimated number of intelligent civilizations in our galaxy, combined with the potential solutions to the Fermi Paradox, suggest that ancient aliens may have visited Earth in the past, influencing human history and leaving behind clues for us to uncover.

The Search for Extraterrestrial Intelligence (SETI) has played a pivotal role in the ongoing quest to find and communicate with intelligent alien life since its inception in the 1960s. Traditional SETI efforts have primarily concentrated on detecting radio signals from extraterrestrial civilizations, based on the assumption that they would use communication technologies similar to those employed by humans. However, in recent years, advancements in astronomy and technology have broadened the scope of SETI initiatives, paving the way for the exploration of innovative detection methods that may provide further evidence to support the existence of ancient aliens.

One such emerging detection method is the search for optical and infrared signals from extraterrestrial civilizations. Optical SETI focuses on detecting brief, intense pulses of light that could indicate the use of powerful lasers as a form of interstellar communication. Infrared SETI, on the other hand, seeks to identify the heat signatures produced by advanced civilizations' energy consumption, which could be detectable over vast interstellar distances.

These alternative methods for searching for extraterrestrial signals could potentially uncover evidence of ancient alien civilizations that utilized more advanced communication technologies than those currently employed by humans. The detection of optical or infrared signals would not only provide compelling evidence for the existence of extraterrestrial life but also reinforce the notion that ancient aliens may have been technologically advanced compared to our ancestors.

Another promising avenue of research is the analysis of exoplanetary atmospheres for signs of artificial pollutants or other indicators of advanced civilizations. As our ability to detect and study exoplanets continues to improve, scientists are increasingly able to scrutinize the chemical compositions of these distant worlds' atmospheres. The detection of certain gases or chemical compounds, particularly those that are unlikely to occur naturally, could be a sign of industrial activity by extraterrestrial civilizations.

This method of searching for extraterrestrial life has the potential to provide indirect evidence of ancient alien civilizations that may have visited Earth in the past. If a planet hosting an advanced civilization can be identified, it would lend credence to the possibility that such beings could have traveled to our solar system and interacted with early human societies.

Technosignatures, or signs of technology created by extraterrestrial civilizations, represent another potential method for detecting intelligent alien life. These can include a wide variety of indicators, such as artificial structures in space, large-scale engineering projects, or even the waste heat generated by advanced computing systems. By searching for technosignatures, scientists aim to uncover evidence of technological activities that would suggest the presence of intelligent extraterrestrial civilizations.

The discovery of technosignatures could provide crucial support for the Ancient Alien Theory, as it would demonstrate the existence of advanced civilizations capable of undertaking the monumental engineering feats and sharing the knowledge attributed to ancient aliens by proponents of the theory.

The ongoing search for extraterrestrial intelligence, driven by SETI and its expanding range of detection methods, holds the potential to uncover evidence that may support the existence of ancient aliens. As technology advances and our understanding of the cosmos deepens, the possibility of finding intelligent extraterrestrial life becomes increasingly plausible, potentially reshaping our understanding of human history and our place in the universe.

The groundbreaking discovery of exoplanets has opened new avenues of research and greatly expanded our understanding of the potential for extraterrestrial life. The revelation that thousands of planets orbit stars beyond our solar system raises the possibility that some of these worlds might be home to intelligent beings capable of interstellar travel and interaction with ancient human civilizations.

The habitable zone, often referred to as the "Goldilocks zone," is the region around a star where conditions are neither too hot nor too cold for liquid water to exist on a planet's surface. Liquid water is a crucial ingredient for life as we know it, providing a medium for complex chemistry and serving as a solvent for a wide variety of biological processes.

The presence of exoplanets within their star's habitable zone increases the likelihood that these worlds may harbor complex life forms, including intelligent species. This bolsters the argument for the existence of ancient aliens, as it suggests that there could be numerous civilizations throughout the galaxy that have evolved to a level of technological sophistication capable of visiting Earth in the distant past.

As our telescopes and detection techniques continue to improve, we are discovering an increasingly diverse array of exoplanets, ranging from small, rocky worlds to massive gas giants. This diversity points to the potential for a wide variety of habitable environments, further increasing the probability of encountering intelligent extraterrestrial life forms.

Some of these exoplanets exhibit conditions that are markedly different from those found on Earth, yet may still be conducive to the development of life. For instance, "super-Earths" – planets with a mass greater than that of Earth but less than that of Neptune – could offer stable environments for life to thrive, with the potential for advanced civilizations to arise.

The discovery of Earth-like exoplanets orbiting within their star's habitable zone lends credence to the Ancient Alien Theory by suggesting that advanced, intelligent life may have developed elsewhere in the cosmos.

These beings may have achieved interstellar travel capabilities, allowing them to visit and interact with ancient human civilizations.

The idea that extraterrestrial beings may have played a role in shaping human history and culture becomes more plausible as we continue to uncover evidence of potentially habitable worlds beyond our solar system. These discoveries provide a scientific foundation for the Ancient Alien Theory, encouraging further investigation into the possibility of ancient encounters between humans and intelligent extraterrestrial life.

As we continue to explore the galaxy and develop new technologies for detecting and studying exoplanets, our understanding of the potential for intelligent extraterrestrial life will continue to evolve. Future space missions, such as the James Webb Space Telescope and next-generation observatories, will allow us to probe the atmospheres of these distant worlds for signs of life, including potential biomarkers and technosignatures indicative of advanced civilizations.

By deepening our understanding of exoplanets and their potential habitability, we may one day uncover definitive evidence that supports the Ancient Alien Theory, shedding light on our shared cosmic history and revealing the true extent of intelligent life in the universe.

The study of extremophiles has been pivotal in broadening our understanding of life's resilience and adaptability, which in turn has implications for the existence of ancient aliens. Extremophiles, organisms that can thrive in extreme environments, have been discovered in various inhospitable conditions on Earth. For instance, extremophiles known as thermophiles and hyperthermophiles can survive in high-temperature environments like deep-sea hydrothermal vents and volcanic hot springs. Psychrophiles, on the other hand, can endure extremely cold environments, such as the Arctic and Antarctic ice.

Other extremophiles include acidophiles, which thrive in highly acidic conditions, and halophiles, which require high salt concentrations for survival. Additionally, some extremophiles, such as radiotolerant organisms, can withstand high levels of radiation, like those found in the radioactive cooling pools of nuclear reactors.

These remarkable findings imply that life could potentially exist in extreme environments on other planets or moons, which might be vastly different from Earth-like conditions. For instance, the discovery of extremophiles in Earth's subsurface has led scientists to speculate that similar life forms could exist beneath the icy crusts of Jupiter's moon Europa or Saturn's moon Enceladus, both of which are believed to harbor subsurface oceans.

By demonstrating that life can persist in a wide array of environments, the study of extremophiles lends support to the notion that intelligent extraterrestrial life forms could have evolved in diverse celestial settings. These extraterrestrial beings may have developed advanced technologies and capabilities to overcome their harsh environments, potentially leading them to explore other planets, including Earth, in search of resources or knowledge.

Additionally, researchers have proposed the possibility of alternative biochemistries that could support life based on different molecular structures or elements. For example, while Earth-based life relies on carbon as the primary element for building complex molecules, it has been suggested that extraterrestrial life forms might be based on silicon or other elements. Silicon, like carbon, can form four chemical bonds and create stable, complex molecules.

Boron, nitrogen, and phosphorus have also been suggested as potential alternative elements for supporting life. The existence of alternative biochemistries would significantly expand the range of potential habitats for life in the universe, thus increasing the likelihood of encountering intelligent extraterrestrial beings.

The study of extremophiles and the exploration of alternative biochemistries not only provide fascinating insights into the adaptability and diversity of life but also support the argument for the existence of ancient aliens. If life can thrive in an array of extreme environments and develop unique biochemistries, it becomes increasingly plausible that ancient alien civilizations could have arisen and visited Earth in the distant past, influencing the course of human history and culture.

Panspermia, a hypothesis that suggests life on Earth may have originated from microbial life forms transported from other celestial bodies, has gained traction in recent years due to mounting scientific evidence. This concept not only expands our understanding of the possible origins of life but also strengthens the argument for the existence of intelligent extraterrestrial life and its potential interaction with ancient human civilizations.

The discovery of organic molecules and amino acids, the fundamental building blocks of life, in meteorites and on comets lends support to the panspermia hypothesis. For example, the Murchison meteorite, which landed in Australia in 1969, contained over 90 amino acids, including some not found on Earth. Similarly, the European Space Agency's Rosetta mission detected the amino acid glycine and other organic molecules on the surface of Comet 67P/Churyumov-Gerasimenko.

These findings indicate that the essential components for life can form in space and potentially be delivered to planets via celestial bodies, such as comets and asteroids.

In addition to the discoveries in meteorites and comets, laboratory simulations have demonstrated that simple organic molecules can form under conditions that mimic those found in space. Experiments such as the Miller-Urey experiment and subsequent studies have shown that by simulating the conditions of early Earth or interstellar environments, amino acids and other organic compounds can be synthesized.

These results strengthen the case for panspermia by showing that the basic ingredients for life can originate in extraterrestrial environments and potentially seed life on habitable planets, including Earth.

The resilience of certain microscopic organisms, such as tardigrades, or "water bears," further supports the possibility of panspermia. Tardigrades are known for their ability to survive extreme conditions, including the vacuum of space, extreme temperatures, and high levels of radiation. Their extraordinary survivability raises the possibility that microbial life could survive the harsh journey through space on a celestial body and subsequently thrive upon reaching a habitable environment.

If panspermia played a role in the origins of life on Earth, it is plausible to consider that it could also have contributed to the development of life on other planets. As our knowledge of exoplanets and their habitable zones grows, the likelihood of finding planets with similar conditions to Earth increases. Consequently, it is possible that panspermia could have seeded life on these planets, leading to the emergence of intelligent extraterrestrial life.

The panspermia hypothesis and its potential role in the development of intelligent extraterrestrial life have profound implications for the Ancient Alien Theory. If extraterrestrial beings emerged on other planets through a process similar to that which occurred on Earth, it is conceivable that they could have developed advanced technology and knowledge, which they may have shared with ancient human civilizations during past interactions.

Evidence of these interactions can be found in various ancient texts, artifacts, and structures that have perplexed researchers and fueled speculation about the role of ancient aliens in human history.

The panspermia hypothesis provides compelling support for the existence of intelligent extraterrestrial life and its potential interactions with ancient human civilizations. The discovery of organic molecules and amino acids in space, along with laboratory simulations and the resilience of microbial life, demonstrates the plausibility of life originating beyond Earth. As our understanding of the universe expands, the possibility of uncovering further evidence of extraterrestrial life and its impact on our past becomes increasingly likely. This, in turn, reinforces the Ancient Alien Theory, inviting us to reevaluate human history and consider the possibility that our ancestors may have had contact with intelligent beings from other worlds.

The concept of panspermia, in conjunction with the increasing number of exoplanet discoveries, is driving scientific research towards the search for extraterrestrial life. Future missions, such as NASA's James Webb Space Telescope and the European Space Agency's PLATO mission, will focus on detecting biomarkers in the atmospheres of exoplanets, which could provide further evidence for the existence of life beyond Earth.

As we continue to explore the cosmos and uncover new information about the potential for life in the universe, the Ancient Alien Theory and its implications will undoubtedly remain a topic of great interest and debate. The possibility that ancient human civilizations may have had contact with intelligent extraterrestrial beings not only challenges our current understanding of human history but also invites us to consider the broader implications of such interactions for our future as a species.

Ultimately, the panspermia hypothesis and the associated search for intelligent extraterrestrial life serve to remind us of the vastness and complexity of the universe. As we continue to explore and learn more about the cosmos, we may one day uncover definitive evidence of ancient alien interactions, opening the door to a deeper understanding of our past and the potential for future collaboration with other intelligent life forms.

The Ancient Alien Theory suggests that extraterrestrial beings visited Earth in the distant past and influenced the development of human civilization. Proponents of this theory argue that evidence of these encounters can be found in various ancient texts, artworks, and structures.

One of the earliest known civilizations, the Sumerians, documented their history and beliefs in cuneiform texts. Among these texts are accounts of beings called the Anunnaki, who were said to have descended from the sky. The Anunnaki were described as powerful, god-like figures who were involved in the creation of humanity and the development of civilization. Proponents of the Ancient Alien Theory argue that the Anunnaki could be interpreted as extraterrestrial beings who visited Earth and shared their advanced knowledge with the Sumerians.

The Nazca Lines are a series of massive geoglyphs etched into the desert landscape of southern Peru. These intricate designs, which can only be fully appreciated from an aerial perspective, include geometric patterns, animals, and human-like figures. Some proponents of the Ancient Alien Theory believe that the Nazca Lines were created as landing sites or navigational markers for extraterrestrial spacecraft. They argue that the scale and complexity of the geoglyphs, as well as the necessity for an aerial perspective, suggest a level of technological sophistication beyond what the Nazca people could have achieved on their own.

The ancient Indian epic, the Mahabharata, is one of the most extensive literary works in human history. It contains numerous stories and descriptions that some believe could be interpreted as encounters with extraterrestrial beings and their advanced technology. One example is the vimana, a flying machine mentioned throughout the epic. Vimanas were described as being capable of traveling vast distances, hovering, and even engaging in aerial combat. Proponents of the Ancient Alien Theory argue that these descriptions are remarkably similar to modern accounts of UFOs and could represent ancient encounters with extraterrestrial spacecraft.

The construction of the Egyptian pyramids and the Sphinx remains an enduring mystery. The sheer size, precision, and alignment of these structures have led some proponents of the Ancient Alien Theory to suggest that they were built with the help of extraterrestrial technology. They argue that the ancient Egyptians may have received knowledge and assistance from visiting aliens, enabling them to construct these monumental works.

The ancient site of Puma Punku in Bolivia features remarkable stonework, including finely cut and precisely fitted stones weighing several tons. The level of precision and craftsmanship displayed at Puma Punku has led some proponents of the Ancient Alien Theory to argue that the builders must have had access to advanced technology, potentially provided by extraterrestrial visitors.

Proponents of the Ancient Alien Theory assert that various ancient texts, artworks, and structures provide evidence of extraterrestrial visitation in the past. While skeptics argue that these examples can be explained through cultural, religious, or mythological interpretations, advocates of the theory maintain that these accounts should be considered as potential evidence of past interactions between humans and intelligent extraterrestrial beings. As we continue to explore our past and unravel the mysteries of human history, the possibility of ancient alien encounters remains an intriguing and thought-provoking concept.

Unidentified flying objects (UFOs) and alien abduction stories have played a prominent role in popular culture and fueled the imagination of Ancient Alien Theory proponents. These phenomena, spanning across centuries and even dating back to ancient times, have provided tantalizing hints of possible extraterrestrial visitation and interactions with humans.

Throughout history, numerous accounts of strange aerial phenomena have been documented in ancient texts, art, and oral traditions. For example, in the Bible, the prophet Ezekiel describes a vision of a "wheel within a wheel" that descended from the sky, an account that some interpret as an encounter with an extraterrestrial spacecraft. Similarly, ancient Indian texts such as the Mahabharata and the Ramayana depict flying chariots and powerful weapons that resemble modern descriptions of UFOs and advanced technology.

These ancient accounts provide intriguing evidence that supports the Ancient Alien Theory, suggesting that our ancestors may have witnessed extraterrestrial vehicles and possibly even interacted with their occupants.

In the modern era, UFO sightings have been reported by people from all walks of life, including pilots, military personnel, and ordinary citizens. Some of these sightings have been corroborated by radar data, photographs, and videos, making them difficult to dismiss as mere misinterpretations or hoaxes.

Governments around the world have taken an interest in these phenomena, conducting investigations and even releasing previously classified UFO documents. Notably, the U.S. Department of Defense's Advanced Aerospace Threat Identification Program (AATIP) studied unexplained aerial phenomena, indicating that governments take these sightings seriously and acknowledge the possibility of extraterrestrial origins.

Alongside UFO sightings, numerous individuals have reported experiencing alien abductions, during which they claim to have been taken aboard extraterrestrial spacecraft and subjected to various examinations or procedures. Some common themes in these accounts include encounters with humanoid beings, the presence of advanced technology, and descriptions of the interior of the spacecraft.

Interestingly, many abduction experiencers recount details that bear striking similarities to ancient depictions of extraterrestrial beings and their technology. For example, the alleged abductors are often described as tall, humanoid creatures with elongated skulls – a feature that has been observed in ancient Egyptian art and the Paracas skulls found in Peru.

Some alien abduction experiencers report physical evidence of their encounters, such as unexplained scars, implants, or even pregnancies. Although skeptics argue that these phenomena could be attributed to mundane causes or psychological factors, the consistency and specificity of these cases warrant further investigation.

In addition to physical evidence, a number of abduction experiencers undergo hypnosis or regression therapy to recover memories of their encounters. These sessions often reveal additional details and corroborate their conscious recollections, further supporting the notion that these experiences may have a basis in reality and could be evidence of extraterrestrial interaction.

The UFO phenomenon and alien abduction experiences, both in ancient and modern times, provide compelling evidence that favors the existence of ancient aliens. The similarities between modern accounts and ancient descriptions of extraterrestrial beings and their technology suggest that our ancestors may have had encounters with advanced civilizations from other worlds. As our understanding of these phenomena continues to evolve, the possibility of uncovering definitive evidence of extraterrestrial visitation and interaction with humans becomes increasingly likely.

The discovery of intelligent extraterrestrial life would have far-reaching consequences, forcing us to reevaluate our understanding of our place in the universe. This revelation would challenge established scientific paradigms, religious beliefs, and societal structures, while also providing invaluable insights into the nature of life and intelligence itself.

The confirmation of intelligent extraterrestrial life would force scientists to reassess many of the theories and assumptions that underpin our current understanding of the cosmos. It could lead to new insights into the origins and evolution of life, the nature of intelligence and consciousness, and the development of advanced technologies. Furthermore, the discovery of ancient aliens would validate the Ancient Alien Theory, demonstrating that extraterrestrial beings have influenced human civilization throughout history.

The existence of intelligent extraterrestrial life would significantly impact the world's religious beliefs and spiritual practices. It would call into question the uniqueness of human beings and our relationship with the divine. While some religious perspectives may struggle to reconcile the existence of extraterrestrial beings with their doctrine, others may adapt and find new ways to incorporate these beings into their belief systems. The discovery of ancient aliens could also lend support to the notion that gods or divine beings from various mythologies were, in fact, visitors from other worlds.

The implications of discovering intelligent extraterrestrial life would extend beyond science and religion, reshaping our societal structures and the way we perceive ourselves as a species. This revelation could lead to a greater sense of global unity, as humanity recognizes its shared identity in the face of a vast and diverse cosmos. The knowledge that ancient aliens may have played a role in our development could inspire a reevaluation of our cultural heritage, fostering greater understanding and appreciation of the interconnectedness of human history.

Interaction with intelligent extraterrestrial life could yield unprecedented opportunities for scientific and technological advancement. By studying the technology and knowledge of ancient aliens, humanity could gain insights into alternative energy sources, advanced materials, and novel forms of propulsion, potentially revolutionizing our understanding of the universe and our ability to explore it. Furthermore, the collaboration between human and extraterrestrial civilizations could facilitate the exchange of ideas and knowledge, leading to breakthroughs in medicine, agriculture, and other fields essential to human well-being.

The discovery of intelligent extraterrestrial life, particularly ancient aliens, would raise a host of ethical and philosophical questions. How should we treat these beings, and what rights and responsibilities do we have toward them? What are the moral implications of past interactions between ancient human civilizations and extraterrestrial beings? These questions would likely lead to a reexamination of our ethical frameworks and an expansion of our moral considerations to include non-human intelligences.

In conclusion, the impact of discovering intelligent extraterrestrial life, especially ancient aliens, would be monumental, challenging our scientific, religious, and societal foundations. As our understanding of the cosmos continues to grow and our search for extraterrestrial life advances, the prospect of uncovering evidence of intelligent life beyond Earth and its influence on our past becomes increasingly likely. The implications of this discovery would reverberate throughout human culture and history, forever transforming our perception of ourselves and our place in the universe.

Chapter 4

Tradition and Myth

Beliefs of Ancient Alien Contact in History and Culture

Throughout history, cultures from around the world have shared myths and traditions that suggest contact with extraterrestrial beings. These accounts often describe otherworldly visitors who descended from the heavens, imparting wisdom and technology to our ancestors. This chapter will examine these stories, highlighting the striking similarities across cultures and their potential implications for the Ancient Alien Theory.

The Sumerian civilization, which thrived in Mesopotamia (modern-day Iraq) around 4500 BCE, is renowned for its many contributions to human history, such as the development of the first known system of writing, the invention of the wheel, and the establishment of sophisticated systems of governance and law. However, one of the most intriguing aspects of Sumerian culture is its mythology, particularly the accounts of the Anunnaki, a group of powerful deities who are believed to have descended from the heavens.

Sumerian texts, inscribed on cuneiform tablets, contain numerous references to the Anunnaki, who were considered to be responsible for the creation of humanity and the shaping of civilization. These texts recount how the Anunnaki came to Earth in search of resources and how they created human beings to serve as their labor force. The Anunnaki were said to have bestowed upon humans the knowledge necessary to cultivate crops, build cities, and develop complex systems of governance.

In Sumerian mythology, the Anunnaki are often depicted as powerful, divine beings with a vast array of abilities and knowledge. They are described as having immense physical stature, advanced technology, and profound wisdom. Some researchers have interpreted these descriptions as evidence of extraterrestrial beings who visited Earth during ancient times, sharing their advanced knowledge and technology with early human civilizations.

Proponents of the Ancient Alien Theory argue that the Sumerian accounts of the Anunnaki provide compelling evidence of ancient extraterrestrial contact. They point to several key aspects of these accounts that suggest the Anunnaki were not merely mythological constructs but, rather, real beings who possessed advanced knowledge and technology beyond the capabilities of the Sumerians.

For example, the Sumerian texts describe how the Anunnaki used advanced genetic engineering techniques to create human beings in their image. They also recount how the Anunnaki traveled between Earth and their home planet using sophisticated spacecraft, a concept that aligns with modern understandings of interstellar travel. Additionally, the Anunnaki were said to have possessed advanced knowledge of astronomy, including information about the existence of planets in our solar system that was only confirmed by modern science thousands of years later.

One of the most compelling arguments in favor of the Ancient Alien Theory is the prevalence of similar stories across various cultures around the world. Many ancient civilizations, such as the Egyptians, the Maya, and the Native Americans, have myths and legends that describe powerful beings who descended from the heavens and shared their advanced knowledge with humanity. These cross-cultural similarities suggest that these stories may be based on real encounters with extraterrestrial beings who visited Earth in ancient times.

In the case of the Sumerians and the Anunnaki, the remarkable level of detail in their accounts, as well as the advanced knowledge attributed to these beings, lends further credence to the possibility that they were, in fact, ancient astronauts. While critics argue that these stories can be explained as mythological constructs or the result of cultural contamination, proponents of the Ancient Alien Theory maintain that the evidence for ancient extraterrestrial contact is too compelling to dismiss.

The Sumerian accounts of the Anunnaki provide a fascinating and persuasive example of potential ancient alien contact. The advanced knowledge and abilities attributed to these beings, as well as the striking similarities between their stories and those of other ancient cultures, suggest that they may have been real extraterrestrial visitors who played a crucial role in shaping the course of human history. As we continue to explore the mysteries of our past and the cosmos, the possibility of ancient alien contact remains an important area of inquiry and a tantalizing prospect for understanding our place in the universe.

The intriguing accounts of the Anunnaki have captured the imagination of researchers, historians, and enthusiasts, inspiring a wealth of books, documentaries, and even popular films that explore the possible connections between these ancient beings and extraterrestrial visitors. These accounts have also influenced modern theories about the origins of human civilization and the development of advanced technologies in ancient times.

As we continue to uncover new information about our past and explore the cosmos, the Anunnaki's role in human history raises important questions about the nature of human development, the origins of our knowledge and technology, and the potential for life beyond Earth. By examining these accounts and considering their implications, we can gain a deeper understanding of the broader story of human civilization and our place in the cosmos.

The potential existence of ancient aliens, such as the Anunnaki, presents an exciting area for future research and exploration. As archaeological and scientific techniques continue to advance, we may be able to uncover more concrete evidence that supports or refutes the Ancient Alien Theory. The discovery of new artifacts, ancient texts, or even the potential identification of extraterrestrial DNA in ancient human remains could provide critical insights into the possible interactions between our ancestors and otherworldly beings.

In addition to archaeological and genetic research, ongoing efforts to explore the cosmos and search for extraterrestrial life may also provide valuable context for our understanding of ancient alien contact. The discovery of exoplanets, the continued search for microbial life on Mars, and the development of new technologies for interstellar communication all have the potential to shed new light on the possibility of ancient alien contact and the broader implications for our understanding of life in the universe.

As we continue to delve into the mysteries of our past and explore the vast reaches of the cosmos, the stories of the Anunnaki and other ancient alien encounters serve as a powerful reminder of the infinite possibilities that await us in our ongoing quest for knowledge and understanding. By remaining open to the potential existence of ancient aliens and engaging in rigorous, open-minded inquiry, we can further expand our horizons and deepen our appreciation for the rich tapestry of human experience and the uncharted realms of the cosmos.

The ancient Egyptians, known for their remarkable achievements in architecture, mathematics, and astronomy, left behind intriguing evidence suggesting possible ancient alien contact. Egyptian mythology is filled with references to gods associated with the stars and the heavens, such as the god Ra, who was believed to have descended from the sky in a celestial boat. The mythology surrounding the construction of the Great Pyramid of Giza has also been linked to the Ancient Alien Theory, with proponents arguing that the evidence for extraterrestrial involvement in the construction of the pyramids is too compelling to ignore.

Egyptian gods were often depicted as having celestial origins or connections. For instance, the god Osiris, who was associated with the afterlife, was believed to reside in the constellation Orion. Similarly, the goddess Isis, the wife of Osiris, was linked to the star Sirius, which was considered the "soul" of Isis. These associations with celestial bodies suggest that the ancient Egyptians may have been influenced by extraterrestrial beings who visited Earth in the past.

The construction of the Great Pyramid of Giza remains one of the most impressive feats of engineering in human history. According to some interpretations, the ancient Egyptians may have received assistance from extraterrestrial visitors in building this massive structure.

The Great Pyramid's alignment with the stars in Orion's Belt is a prime example of the advanced astronomical knowledge that the ancient Egyptians possessed. This precise alignment, along with the sophisticated internal architecture of the pyramid, has led some researchers to speculate that it was designed with the help of advanced knowledge and technology beyond the capabilities of the ancient Egyptians. The construction process, which involved moving and placing massive stone blocks with incredible precision, could also indicate the use of advanced technology or techniques that were not available to the Egyptians at the time.

The mysterious shafts within the Great Pyramid, which point towards specific stars, further support the idea of extraterrestrial involvement. These shafts, once believed to be air vents, are now thought to have been constructed with a specific purpose: to align with certain stars at specific times. This level of precision in the pyramid's design and construction implies a deep understanding of astronomy that seems unlikely for the ancient Egyptians to have achieved on their own.

Egyptian knowledge of mathematics and astronomy was well ahead of its time. For example, they had a surprisingly accurate understanding of the Earth's axial tilt and the length of the solar year. Additionally, they were able to predict solar and lunar eclipses with great accuracy, a feat that would have required an advanced understanding of celestial mechanics. The complexity of their knowledge in these areas suggests that they may have received guidance from beings with a more advanced understanding of the cosmos.

Critics argue that the Egyptians' achievements can be attributed to their own ingenuity and innovation. However, proponents of the Ancient Alien Theory maintain that the level of sophistication in their knowledge, combined with the myths surrounding their gods and celestial connections, makes a compelling case for ancient alien contact.

The intriguing evidence found in ancient Egyptian mythology, the construction of the Great Pyramid of Giza, and their advanced knowledge of mathematics and astronomy lends support to the Ancient Alien Theory. While skeptics continue to question these claims, the enduring mysteries surrounding the ancient Egyptians and their accomplishments continue to fuel speculation about the possible existence of extraterrestrial involvement in our past.

The ancient texts of Hinduism, including the Vedas and the Mahabharata, offer fascinating insights into the mythological world of ancient India, with detailed accounts of flying machines called Vimanas. These vehicles were often associated with the gods and were described as being powered by advanced technology, capable of traversing vast distances through the air, underwater, and even in space.

The descriptions of Vimanas in these texts are remarkably detailed, providing information on their construction, propulsion systems, and capabilities. For example, the ancient Sanskrit text called the Vaimanika Shastra, which is believed to be a compilation of even older texts, provides comprehensive details on the construction, operation, and use of Vimanas. The text also describes different types of Vimanas, such as the Rukma Vimana, which was said to be shaped like a double-decked, circular aircraft with portholes and a dome.

Some researchers have suggested that these accounts are evidence of ancient alien technology, as the concepts and descriptions align with modern understandings of aerospace engineering and interstellar travel. The propulsion systems described in the texts, such as the use of mercury vortex engines and other advanced technologies, share striking similarities with concepts currently being explored by scientists and engineers working in the field of space exploration.

In addition to the technological aspects, the ancient texts also describe the weaponry used on Vimanas, including powerful energy weapons and even nuclear-like bombs. One such example can be found in the Mahabharata, which recounts a devastating war involving Vimanas, with descriptions of weapons that could annihilate entire cities and cause widespread destruction, eerily reminiscent of modern-day nuclear weapons.

Proponents of the Ancient Alien Theory argue that the specificity and consistency of these accounts suggest that they may be based on real encounters with advanced extraterrestrial technology. The notion that ancient India was visited by extraterrestrial beings who shared their advanced knowledge and technology with humans could help explain the rapid advancements in science, architecture, and astronomy that characterized this period in Indian history.

Furthermore, the depiction of gods and celestial beings piloting these advanced flying machines suggests a cultural exchange between ancient humans and extraterrestrial visitors. The fact that these beings were often depicted as teachers and benefactors to humanity, sharing their wisdom and knowledge, is consistent with the pattern observed in other ancient mythologies that are believed to describe encounters with ancient aliens.

Critics argue that the Vimanas are purely mythological constructs, representing the ancient Hindus' imaginative ideas about the capabilities of their gods. They maintain that the detailed descriptions of Vimanas and their technology may be the result of creative storytelling rather than factual accounts of real events.

However, proponents of the Ancient Alien Theory maintain that the level of detail and the striking similarities between the Vimana descriptions and modern aerospace technology are too compelling to dismiss as mere coincidence or imagination. They argue that the consistency of these accounts, along with the prevalence of similar stories in other cultures, suggests that they may be based on real encounters with advanced extraterrestrial technology.

The detailed accounts of Vimanas found in ancient Hindu texts provide a strong case for the possibility of ancient alien contact in India. The descriptions of advanced technology and the striking similarities with modern aerospace concepts suggest that these accounts may be more than just mythological tales, but rather evidence of encounters with extraterrestrial beings who shared their knowledge with our ancestors. By examining these accounts and their potential implications for human history, we can gain a deeper understanding of our past and the possible existence of advanced extraterrestrial civilizations.

Many Native American cultures across North America maintain oral traditions that recount interactions with the Star People, otherworldly beings who visited Earth from the cosmos. These legends often portray the Star People as wise teachers who imparted crucial knowledge to indigenous peoples, helping to shape their societies and cultures. Proponents of the Ancient Alien Theory argue that the prevalence of these tales across diverse Native American tribes may indicate actual encounters with extraterrestrial beings.

The Hopi, a Native American tribe from the southwestern United States, possess an extensive mythology surrounding the Star People, whom they also refer to as Kachinas. According to Hopi legend, these celestial beings played a pivotal role in human creation and the development of their culture. The Kachinas are believed to have provided the Hopi with essential knowledge in agriculture, astronomy, and the natural world.

One fascinating aspect of Hopi mythology is the story of the Ant People, who are believed to have saved the Hopi during two global cataclysms. According to the Hopi Emergence Myth, the Ant People sheltered the Hopi underground, teaching them methods of food storage and survival during these times of crisis. The advanced knowledge that the Ant People shared with the Hopi is seen by some as evidence of ancient alien contact.

The Lakota Sioux, a Native American tribe from the Great Plains, have their own rich tradition of Star People stories. In their legends, the Star People, known as Wicahpi Oyate, descended from the Pleiades star cluster to impart wisdom and spiritual knowledge to their ancestors. The Lakota Sioux still hold annual ceremonies to honor the Wicahpi Oyate and the knowledge they shared.

The Iroquois Confederacy, a group of Native American tribes in the northeastern United States and eastern Canada, has its own intriguing oral tradition involving a celestial visitor. According to their creation story, the Sky Woman descended from the heavens and played a crucial role in the formation of the Earth. The Sky Woman's origin, coupled with her importance in the creation narrative, has led some to speculate that she may represent an ancient alien visitor.

While critics argue that Star People stories can be explained as mythological constructs, reflecting Native Americans' efforts to comprehend the world around them, proponents of the Ancient Alien Theory point to the striking similarities across diverse cultures as evidence of real extraterrestrial encounters. The common themes of celestial origins, advanced knowledge, and the crucial role these beings played in the development of Native American cultures suggest that these stories may be rooted in actual experiences with visitors from other worlds.

As we continue to explore the depths of human history and the vast reaches of the cosmos, the legends of the Star People remain an intriguing area of inquiry. By examining these narratives, we can gain valuable insights into the potential connections between our ancestors and extraterrestrial beings, and better understand the impact of these encounters on the development of Native American cultures and the broader human experience.

The ancient Mesoamerican civilizations, including the Maya, Aztec, and Olmec, offer fascinating evidence that suggests contact with extraterrestrial beings. One of the most iconic figures in Mesoamerican mythology is the Feathered Serpent, known as Quetzalcoatl to the Aztecs and Kukulkan to the Maya. This enigmatic figure is often depicted as a wise and benevolent god who descended from the heavens and taught humans the arts of agriculture, writing, and other advanced skills.

Some researchers have proposed that the Feathered Serpent may be an ancient astronaut who visited Earth in the remote past, sharing advanced knowledge and technology with the Mesoamerican peoples. This theory gains credence when considering the impressive accomplishments of these civilizations, such as the construction of intricate temple complexes, the development of sophisticated mathematical and astronomical systems, and the creation of elaborate written languages.

For example, the ancient Maya were capable of predicting solar and lunar eclipses, as well as the movements of the planets, with remarkable accuracy. They also developed an advanced calendar system that accurately measured time over vast cycles. It is difficult to imagine how such complex knowledge could have been acquired without the assistance of an external, advanced source of information.

Moreover, the Feathered Serpent is not an isolated figure in Mesoamerican mythology. There are other deities, such as the Aztec god Tlaloc and the Maya rain god Chaac, who are also associated with advanced knowledge and technology. These deities were often depicted with features that could be interpreted as indicative of extraterrestrial origin, such as elongated skulls, large eyes, and unusual attire.

Proponents of the Ancient Alien Theory point out that the striking similarities between the Feathered Serpent and other mythological figures from around the world suggest that these stories may be based on real encounters with advanced extraterrestrial beings. For instance, the ancient Sumerian deity Enki, the Greek god Prometheus, and the Chinese deity Fuxi all share characteristics with the Feathered Serpent. They were all considered to be wise teachers who bestowed advanced knowledge and technology upon humanity.

Another intriguing aspect of the Feathered Serpent legend is the recurring theme of a cataclysmic event, often described as a flood or deluge, which destroyed a previous civilization. The Feathered Serpent is said to have arrived after this event to help humanity rebuild and start anew. This narrative parallels similar stories found in cultures around the world, including the biblical story of Noah and the Sumerian account of the Great Flood. This raises the possibility that the Feathered Serpent and other ancient gods might be ancient astronauts who assisted humanity in recovering from these devastating events.

Critics argue that the Feathered Serpent can be explained as a purely mythological figure, representing the Mesoamericans' attempts to understand the natural world and its cycles. However, proponents of the Ancient Alien Theory contend that the level of detail and the striking similarities between the Feathered Serpent and other mythological figures from around the world point towards real encounters with advanced extraterrestrial beings. The impressive accomplishments of the Mesoamerican civilizations, as well as the recurring themes in their myths and legends, offer tantalizing evidence that supports the existence of ancient aliens and their influence on human history.

The Dogon people of Mali, in West Africa, have a unique and fascinating mythology that centers around the star system Sirius. This indigenous tribe, living in a remote region with a deeply rooted cultural heritage, has managed to preserve their ancestral beliefs and astronomical knowledge for generations. According to Dogon tradition, their ancestors were visited by a race of extraterrestrial beings called the Nommo, who hailed from the Sirius star system. The Nommo were said to be amphibious beings, possessing both humanoid and aquatic characteristics, which may indicate their adaptability to diverse environments in their extraterrestrial home.

The Dogon's knowledge of Sirius B is particularly compelling, as this information was only confirmed by modern astronomers in the 20th century, specifically in 1862 when Sirius B was first observed by telescope. The Dogon also possess knowledge of Sirius C, a hypothetical star in the Sirius star system that remains unconfirmed by modern astronomers but could be proven to exist in the future. Some researchers have argued that the Dogon's advanced astronomical knowledge could only have come from contact with extraterrestrial beings who possessed advanced technology and knowledge of the cosmos.

The fact that the Dogon's knowledge of the Sirius star system predates Western astronomical discoveries by centuries is not easily dismissed. Their descriptions of the Nommo's appearance, the celestial events surrounding Sirius, and the orbital patterns of its stars are surprisingly accurate, given the limited technology available to them. Additionally, the Dogon have a detailed creation myth that revolves around the Nommo, who are said to have played a significant role in the creation of humanity and the establishment of their culture.

The Dogon describe the Nommo as having arrived on Earth in an "ark," which descended from the sky and landed on the ground with great noise and wind. This account has striking similarities to modern descriptions of spacecraft and could be interpreted as evidence of an actual visitation from an advanced extraterrestrial civilization. The Nommo are said to have shared advanced knowledge with the Dogon, including information about agriculture, metallurgy, and other advanced skills, which may have contributed to the development of their society.

Critics maintain that the Dogon's knowledge of Sirius B could be the result of cultural contamination or coincidence. They argue that the Dogon might have acquired this knowledge through contact with Western explorers or missionaries who visited their region in the past. However, proponents of the Ancient Alien Theory argue that the specificity of the Dogon's astronomical knowledge and their detailed accounts of the Nommo suggest that these stories may be based on real encounters with extraterrestrial beings.

Furthermore, the Dogon's knowledge of Sirius B is not easily explained by cultural contamination, as the tribe's remote location and cultural isolation make it unlikely that they would have had access to advanced astronomical information from outside sources. The consistency of their knowledge, passed down through generations via oral tradition, also strengthens the argument in favor of ancient alien contact.

The Dogon people's intricate knowledge of the Sirius star system, their descriptions of the Nommo and their advanced skills, and the parallels between their mythology and modern accounts of extraterrestrial encounters provide compelling evidence that supports the Ancient Alien Theory. While it is crucial to approach these claims with a critical eye, the possibility of ancient alien contact with the Dogon people should not be dismissed and warrants further investigation.

The myths and traditions of ancient cultures from around the world offer an abundance of persuasive evidence in support of the Ancient Alien Theory. While skeptics argue that these stories can be explained as purely mythological constructs or the result of cultural contamination, advocates of the theory emphasize the striking similarities and consistency across a diverse range of cultures. These parallels suggest that these narratives may be rooted in genuine encounters with technologically advanced extraterrestrial beings.

As humanity continues to explore the cosmos and uncover the mysteries of our past, the stories of ancient alien contact preserved in our myths and traditions remain a vital area of inquiry. By closely examining these accounts, we can gain a deeper understanding of the potential connections between our ancestors and extraterrestrial visitors. This understanding can shed light on the implications of these encounters for our perception of human history and our place in the cosmos.

The Ancient Alien Theory may provide valuable insights and context for our ongoing investigations as we move forward with advancements in technology. These advancements, such as the discovery of exoplanets and the search for extraterrestrial life, reinforce the importance of considering the possibility of ancient alien contact. The myths and traditions examined in this chapter serve as a testament to the enduring fascination with ancient alien contact, and they raise tantalizing questions about the nature of life beyond Earth.

In conclusion, the examination of myths and traditions across various cultures reveals a wealth of compelling evidence that supports the existence of ancient aliens. While it is crucial to approach these accounts with a critical eye and consider alternative explanations, the striking similarities and consistency found in these stories from around the world suggest that they may indeed be rooted in real encounters with advanced extraterrestrial beings.

As we continue to explore the cosmos and uncover the mysteries of our past, it is essential that we remain open to the possibility of ancient alien contact and consider the implications of these encounters for our understanding of human history, our place in the universe, and our future as a species. By examining the evidence and engaging in open-minded inquiry, we can further expand our knowledge and deepen our appreciation for the rich tapestry of human experience and the vast, unexplored realms of the cosmos.

Ultimately, the exploration of the Ancient Alien Theory and its connections to the myths and traditions of various cultures has the potential to reshape our understanding of the human experience. By considering the possibility that our ancestors may have been influenced by advanced extraterrestrial beings, we can reassess the origins of our own technological achievements and the development of our civilizations. This perspective may also encourage us to ponder the potential future interactions between humans and extraterrestrial life, as well as the ethical and philosophical questions these encounters might raise.

With an ever-growing body of evidence supporting the Ancient Alien Theory, it becomes increasingly important for researchers, historians, and scientists to collaborate in the pursuit of understanding the potential influence of extraterrestrial beings on our past. By continuing to examine ancient myths, artifacts, and structures through the lens of this theory, we can uncover new insights that will shape our understanding of human history and our place in the vast, interconnected cosmos

Chapter 5

Contact Phenomena

Accounts of Alien Encounters and Hybridization

Throughout history, tales of alien encounters and hybridization have captured the imaginations of many. These accounts, ranging from ancient legends to modern-day testimonies, have been interpreted as evidence for the existence of ancient aliens. This chapter will delve into the various accounts of alleged alien contact and hybridization, examining their credibility and exploring the implications of these phenomena on our understanding of ancient extraterrestrial life.

The idea of divine beings descending from the heavens to interact with humans has been a prevalent theme in many ancient cultures' mythologies. These gods, often possessing advanced knowledge and supernatural abilities, are sometimes depicted as engaging in relationships with humans, resulting in the birth of hybrid offspring. Some researchers argue that these myths may have been inspired by actual encounters with extraterrestrial beings, providing valuable insights into the possible existence of ancient aliens.

Sumerian texts, such as the Epic of Gilgamesh, describe the Anunnaki, a group of gods who were said to have created humanity by combining their DNA with that of primitive Earth creatures. The Anunnaki were depicted as possessing advanced knowledge and technology, which they used to shape human civilization. Some researchers argue that the Anunnaki were ancient astronauts who genetically engineered humans to serve as their labor force, pointing to the advanced knowledge of the Sumerians in areas such as astronomy, mathematics, and agriculture as evidence of extraterrestrial influence.

Ancient Egyptian mythology is replete with tales of gods descending from the heavens to interact with humans. The gods were often portrayed as possessing advanced knowledge and technology, such as the ability to fly through the sky in "boats of millions of years." Some ancient alien theorists suggest that these stories may represent encounters with extraterrestrial beings who visited Earth in the distant past. Furthermore, the incredible architectural feats of the ancient Egyptians, such as the construction of the pyramids, have led some to propose that this civilization received assistance from extraterrestrial visitors.

In Mesoamerican mythology, the figure of the Feathered Serpent, known as Quetzalcoatl to the Aztecs and Kukulkan to the Maya, is a prominent deity associated with knowledge, agriculture, and the arts. This god was often depicted as a flying serpent or dragon and was said to have descended from the heavens to teach humans the arts of civilization. Some ancient alien theorists argue that the Feathered Serpent may represent an extraterrestrial being who visited ancient Mesoamerica, sharing advanced knowledge and technology with the indigenous peoples.

In Chinese mythology, the Dragon Kings are powerful deities who control weather, water, and other natural phenomena. They are often depicted as living in underwater palaces and possessing advanced technology, such as the ability to fly through the sky. Some ancient alien theorists suggest that the Dragon Kings may represent extraterrestrial beings who visited ancient China and shared their advanced knowledge with the Chinese civilization. This theory is supported by the fact that ancient Chinese texts contain detailed descriptions of flying machines, known as "celestial chariots," which some argue could represent advanced extraterrestrial technology.

In Norse mythology, the Aesir are a group of gods who reside in Asgard, a celestial realm connected to Earth by the rainbow bridge Bifrost. The Aesir, led by the chief god Odin, are depicted as possessing advanced knowledge, supernatural abilities, and powerful weapons. Some ancient alien theorists propose that the Aesir may represent extraterrestrial beings who visited ancient Scandinavia, sharing their knowledge and technology with the Norse people. This theory is bolstered by the presence of advanced metalworking techniques in ancient Norse culture, as well as descriptions of the gods wielding incredible weapons, such as Thor's hammer Mjölnir, which some argue could represent advanced extraterrestrial technology.

Sumerian mythology, considered one of the oldest known mythological systems, offers a wealth of information that has led some researchers to believe in the existence of ancient aliens. Central to this belief is the enigmatic group of gods known as the Anunnaki, whose role in the creation of humanity has sparked significant interest and debate among proponents of ancient astronaut theory.

The Sumerian civilization, which emerged in the region of Mesopotamia around 4500 BCE, was renowned for its remarkable advancements in the fields of agriculture, architecture, mathematics, and astronomy. The Sumerians documented their knowledge and beliefs in a series of cuneiform texts, which include detailed accounts of their interactions with the Anunnaki.

The Anunnaki were described as powerful, divine beings who descended from the heavens and played a crucial role in the development of human civilization. According to Sumerian texts, the Anunnaki came to Earth in search of gold and other precious resources. To aid in their endeavors, they decided to create a labor force by combining their DNA with that of Earth's primitive inhabitants. The resulting hybrid beings were the first humans, designed to serve the Anunnaki and fulfill their needs.

This creation story has been interpreted by some researchers as evidence that the Anunnaki were, in fact, ancient astronauts – extraterrestrial beings who visited Earth in the distant past and directly influenced the development of human civilization.

The Sumerians' advanced knowledge in various fields, especially astronomy, is difficult to explain without considering the possibility of external intervention. The Anunnaki were said to have imparted this knowledge to the Sumerians, which could indicate that they possessed advanced technologies and information beyond the capabilities of humans at the time.

The Sumerian texts' description of the Anunnaki combining their DNA with that of primitive Earth creatures to create humans suggests a sophisticated understanding of genetics and genetic manipulation. This level of knowledge far surpasses what would be expected from a civilization that existed over 6,000 years ago, further supporting the idea that the Anunnaki were advanced extraterrestrial beings.

The stories of divine beings descending from the heavens and interacting with humans are prevalent in numerous ancient cultures worldwide, such as the Greek, Hindu, and Native American mythologies. This widespread presence of similar themes could indicate that these myths were inspired by actual encounters with ancient astronauts, like the Anunnaki.

Some researchers argue that certain archaeological findings, such as the Sumerian King List, which records a period of rule by demigods with incredibly long lifespans, and the presence of advanced architectural structures like ziggurats, may serve as physical evidence of the Anunnaki's influence on human civilization.

The Sumerian mythology's account of the Anunnaki and their role in the creation of humanity provides intriguing support for the existence of ancient aliens. The advanced knowledge, genetic engineering, similarities with other mythologies, and physical evidence all contribute to the argument that the Anunnaki were extraterrestrial beings who visited Earth in our distant past and had a profound impact on the development of human civilization.

Greek mythology is replete with stories of gods fathering children with mortal women, resulting in the birth of demigods or heroes. These hybrid offspring possessed extraordinary abilities and often played pivotal roles in the myths and legends of ancient Greece. Some ancient alien theorists suggest that these tales might have been based on real encounters between extraterrestrial visitors and humans, with the demigods representing a hybrid species. This expanded section will delve deeper into the connections between Greek mythology and the ancient alien hypothesis, examining specific stories, characters, and potential evidence for extraterrestrial involvement.

One of the most famous examples of a demigod in Greek mythology is Perseus, the son of Zeus and the mortal woman Danaë. According to the myth, Zeus visited Danaë in the form of a golden shower, impregnating her and leading to the birth of Perseus. Ancient alien theorists propose that this story might represent an encounter between an extraterrestrial being (Zeus) and a human woman, with the golden shower symbolizing a more advanced form of artificial insemination.

Heracles, also known as Hercules, is perhaps the most famous of all Greek demigods. The son of Zeus and the mortal woman Alcmene, Heracles was imbued with superhuman strength and endurance. His legendary twelve labors are considered some of the most iconic stories in Greek mythology. Ancient alien theorists suggest that Heracles' extraordinary abilities may have been the result of his hybrid genetics, a combination of human and extraterrestrial DNA, which could be a product of ancient alien intervention.

In the Book of Genesis, the Nephilim are described as the offspring of the "sons of God" and the "daughters of men." Some ancient alien theorists draw parallels between the Nephilim and the demigods of Greek mythology, suggesting that both groups may represent hybrid species resulting from the union of extraterrestrial beings and humans. This connection between ancient Greek myths and biblical accounts lends further credence to the idea that these stories may have been inspired by real encounters with otherworldly beings.

The demigods of Greek mythology often possessed advanced knowledge and abilities that set them apart from ordinary humans. These extraordinary capabilities included superhuman strength, exceptional intelligence, and the ability to communicate with animals or control the elements. Ancient alien theorists argue that these abilities may be evidence of extraterrestrial influence, as the demigods' hybrid genetics could have endowed them with enhanced traits that were beyond the reach of ordinary humans.

The demigods of Greek mythology had a profound impact on ancient Greek society, shaping the culture and serving as role models for generations to come. Their extraordinary abilities and heroic deeds were celebrated in art, literature, and religious practices. Ancient alien theorists contend that the demigods' influence on ancient Greece may be further evidence of extraterrestrial involvement, as these hybrid offspring played a crucial role in shaping human history and culture.

The stories of demigods in Greek mythology, such as Perseus and Heracles, may hold deeper meaning when viewed through the lens of the ancient alien hypothesis. The extraordinary abilities, advanced knowledge, and lasting impact of these hybrid offspring on ancient Greek society suggest that the tales of gods fathering children with mortal women may have been inspired by real encounters between extraterrestrial beings and humans. As we continue to explore

Hindu mythology, as recorded in the ancient Vedic texts, contains numerous descriptions of gods and their advanced technology. These accounts have led some researchers to argue that the Devas, or gods, were extraterrestrial beings who visited Earth and shared their knowledge with humanity. This section will further explore the connections between Hindu mythology, the Devas, and the Vimanas as potential evidence for ancient alien encounters.

The Devas were a group of gods in Hindu mythology known for their extraordinary powers, advanced knowledge, and mastery of technology. Some ancient alien theorists argue that the Devas were actually ancient astronauts who came to Earth and interacted with human civilizations. These theorists point to the similarities between the Devas and gods of other ancient cultures, such as the Sumerian Anunnaki and the Greek pantheon, as evidence that these beings were extraterrestrial in origin.

In addition to their advanced knowledge and abilities, the Devas were also described as having unique physical characteristics that set them apart from humans, such as multiple heads or arms. These distinctive traits have led some researchers to speculate that the Devas may have been an extraterrestrial species with a unique genetic makeup.

The Vedas also contain numerous descriptions of advanced flying vehicles called Vimanas. These vehicles were said to be used by the Devas to travel vast distances, both on Earth and between worlds. Vimanas were often described as having extraordinary capabilities, such as the ability to become invisible, move at incredible speeds, and emit powerful energy weapons.

Ancient alien theorists argue that the detailed accounts of Vimanas found in the Vedas are evidence of advanced spacecraft used by extraterrestrial beings. They point to the striking similarities between these descriptions and modern-day accounts of UFO sightings and advanced aircraft as evidence that Vimanas were technological marvels, far beyond the capabilities of ancient human civilizations.

Many accounts in Hindu mythology involve the Devas imparting their advanced knowledge to humans, particularly in the areas of science, medicine, and architecture. The ancient Indian texts contain remarkably advanced concepts for their time, including descriptions of advanced surgical procedures, the concept of the atom, and the use of complex mathematical formulas.

Ancient alien theorists argue that this advanced knowledge could not have been developed by ancient human civilizations without external assistance. They suggest that the Devas, as extraterrestrial beings, were responsible for transferring their advanced knowledge to humanity, either directly or through the use of advanced technology. This knowledge transfer may have played a crucial role in the development of early human civilizations and the advancement of human knowledge.

Hindu mythology, as recorded in the Vedas, offers a compelling case for the existence of ancient aliens in the form of the Devas and their advanced flying machines, the Vimanas. The extraordinary knowledge and capabilities of these beings, as well as the advanced technology they possessed, suggest that they may have been extraterrestrial visitors who played a pivotal role in shaping human history and culture. As we continue to explore our past and the mysteries of the universe, the potential connections between Hindu mythology and ancient alien encounters serve as a fascinating area of inquiry, offering tantalizing clues about the origins and nature of extraterrestrial life.

The phenomenon of alien contact and hybridization has not only persisted into the modern era but has also gained increased attention and credibility due to the sheer volume of reported encounters and the compelling nature of the testimonies. Numerous individuals from diverse backgrounds claim to have had direct encounters with extraterrestrial beings, often involving abduction, experimentation, and the creation of hybrid offspring. These accounts, when considered collectively, provide a strong case for the existence of ancient aliens and their continued interaction with humanity.

In August 1976, four friends on a camping trip in the Allagash Wilderness Waterway in Maine reported being abducted by extraterrestrial beings. The men claimed that a bright light approached their canoe, and they were subsequently taken aboard a spacecraft. While onboard, they experienced telepathic communication with the beings, who performed medical examinations and collected genetic material from them. The case is particularly compelling due to the consistency of the four men's independent accounts, which were obtained under hypnosis, and the polygraph tests they passed.

In November 1989, Linda Cortile (also known as Linda Napolitano) claimed to have been abducted from her Manhattan apartment by extraterrestrial beings who levitated her through a window and onto a waiting spacecraft. Her case gained prominence when two bodyguards of a high-ranking United Nations diplomat came forward, stating that they had witnessed the event while on duty. The case is notable for the involvement of multiple witnesses, including those with no prior connection to the abductee, which adds credibility to the account.

Dr. David M. Jacobs, a historian and UFO researcher, has conducted extensive research into the alien abduction phenomenon, with a particular focus on the alleged creation of human-alien hybrids. Through interviews and hypnotic regressions, Jacobs has collected numerous accounts from individuals who claim to have been abducted and subjected to reproductive experimentation. Many of these individuals describe meeting their hybrid offspring during subsequent abduction experiences. Jacobs' work provides a comprehensive body of evidence supporting the existence of a hybridization program conducted by extraterrestrial beings.

In September 1994, over 60 children at the Ariel School in Ruwa, Zimbabwe, reported witnessing a UFO landing and the appearance of several small humanoid beings near their playground. The children described telepathic communication with the beings, who conveyed messages about humanity's future and the need to protect the environment. The case is notable for the number of witnesses, their young age, and the consistency of their accounts, which were documented through interviews and drawings.

As the number of reported alien abduction and hybridization experiences has grown, some individuals claim to have discovered or been invited to visit communities where human-alien hybrids live and interact with one another. These communities are said to represent the integration of hybrid beings into human society, with the ultimate goal of fostering understanding and cooperation between humanity and extraterrestrial life. While these claims are difficult to verify, they provide an intriguing possibility for the continued presence of ancient aliens and their influence on our world.

The persistence of modern accounts of alien contact and hybridization, coupled with the compelling nature of the testimonies and the corroborating evidence in many cases, lends significant weight to the ancient alien hypothesis. These encounters suggest that extraterrestrial beings have not only visited Earth in the past but continue to interact with and influence humanity in profound and mysterious ways. As we further explore these phenomena, we may gain deeper insights into the existence of ancient aliens and their ongoing impact on our world.

The Betty and Barney Hill abduction case remains one of the most compelling and well-documented instances of alleged alien contact in history. The couple's detailed account of their experience, combined with the physical evidence and corroborating factors, has led many researchers to argue that their encounter may provide valuable insights into the potential existence of ancient aliens and their interactions with humanity.

On the night of September 19, 1961, Betty and Barney Hill were driving through rural New Hampshire on their way home from a vacation in Canada. As they traveled along the isolated stretch of road, they noticed a bright light in the sky that appeared to be following their car. Intrigued, they stopped to get a better look at the object, which they described as a large, disc-shaped craft with a row of windows and blue-white lights around its perimeter.

Fearing for their safety, the Hills decided to continue driving, but soon realized that they could not outrun the mysterious craft. As the object descended upon their vehicle, the couple reported experiencing an intense buzzing sensation, followed by a sudden onset of amnesia. When they regained consciousness, they found themselves over 35 miles from where they had last remembered being, with two hours of unaccounted time.

In the weeks that followed, Betty began experiencing vivid nightmares of being taken aboard a spacecraft and subjected to medical examinations by non-human beings. The couple sought help from a psychiatrist, Dr. Benjamin Simon, who specialized in hypnotherapy. Under hypnosis, both Betty and Barney independently recounted remarkably similar experiences of being abducted by extraterrestrial beings, who had performed a series of medical procedures on them.

The beings were described as having gray skin, large, wraparound eyes, and small slits for mouths and noses. The couple recalled being led through the craft by these beings, who communicated with them telepathically. The aliens performed various tests, including taking samples of their hair, nails, and skin, as well as probing their bodies with needle-like instruments.

Perhaps the most compelling aspect of the Hills' story is the star map that Betty claimed to have seen aboard the spacecraft. Under hypnosis, she was able to reproduce the map from memory, which depicted a series of interconnected lines and dots representing stars and their relative positions. In 1969, Marjorie Fish, an amateur astronomer, analyzed the map and concluded that it was a remarkably accurate representation of the Zeta Reticuli star system, located approximately 39 light-years from Earth. Fish's findings have been the subject of much debate, but they remain one of the most intriguing aspects of the Hills' abduction story.

In addition to their detailed accounts, the Hills also presented physical evidence supporting their claims. Their car exhibited unusual magnetic properties following the encounter, causing a compass to spin erratically when placed near the vehicle. Furthermore, both Betty and Barney's watches stopped working on the night of the abduction and never functioned again.

The Hills also reported finding mysterious stains and tears on their clothing, which they could not explain. In later years, Betty would discover a series of circular warts on her body, arranged in a pattern that matched the needle marks she recalled from her abduction experience.

The Betty and Barney Hill abduction case remains one of the most significant and well-documented instances of alleged alien contact. Their detailed account, combined with the physical evidence and the star map, provides a compelling case for the existence of ancient aliens and their potential interactions with humanity. As we continue to explore the mysteries of our universe and seek answers to the age-old questions surrounding extraterrestrial life, the Hills' story serves as an important reminder of the possibility that we are not alone in the cosmos.

The Betty and Barney Hill abduction case had a profound impact on the study of UFOs and alien encounters. Their story brought the concept of alien abduction into the public consciousness and helped to legitimize the experiences of countless other individuals who would come forward with similar accounts in the years that followed.

The Hills' case has been extensively investigated and analyzed by researchers, scientists, and UFO enthusiasts alike. Their story has been featured in numerous books, television documentaries, and articles, serving as a touchstone for discussions about the potential existence of extraterrestrial life and its possible interactions with humanity.

As we continue to seek answers to the questions surrounding the existence of ancient aliens and their potential influence on human history, the Betty and Barney Hill abduction case remains a significant piece of the puzzle. The details of their encounter, along with the physical evidence and the star map, provide a compelling argument in favor of the existence of ancient aliens.

By examining cases like the Hills', we can gain valuable insights into the potential nature of extraterrestrial life and its implications for our understanding of the universe. As humanity advances in technology and our knowledge of the cosmos expands, we may one day be able to determine with certainty whether stories like the Hills' represent genuine encounters with beings from other worlds or are simply the product of human imagination and cultural influences. Until then, the possibility of ancient alien contact and the enduring fascination with the unknown will continue to captivate researchers, enthusiasts, and skeptics alike.

The Travis Walton abduction case is one of the most well-known and widely-discussed alien abduction incidents in history. On November 5, 1975, Travis Walton, a logger working in the Apache-Sitgreaves National Forest in Arizona, vanished without a trace after his coworkers witnessed an unidentified flying object (UFO) hovering above their work site. Walton reappeared five days later, disoriented and with an extraordinary tale of abduction by non-human entities. The Walton case offers a wealth of compelling evidence and testimony that lends credibility to the ancient alien theory.

The Walton abduction case is unique in that it involved multiple witnesses who corroborated the UFO sighting. Walton was part of a seven-man logging crew that included his best friend and brother. As the crew was leaving the work site, they saw a bright, glowing object hovering approximately 15 to 20 feet above the ground. Walton, driven by curiosity, approached the object and was subsequently struck by a beam of light, which sent him flying backward.

The other crew members, terrified by what they had just witnessed, fled the scene. However, they returned shortly after to search for Walton, only to find that both he and the UFO had vanished. The crew members later reported the incident to the police and were subjected to rigorous questioning and polygraph tests. All of the tests indicated that the witnesses were telling the truth, strengthening the case for the authenticity of the UFO sighting and Walton's subsequent abduction.

When Walton reappeared five days later, he provided a detailed account of his time aboard the spacecraft. He described waking up on a table inside the craft, surrounded by non-human entities. These beings had large, dark eyes, and their heads were disproportionately large compared to their slender bodies.

Walton recounted that he was subjected to medical examinations and observed various advanced technologies during his time on the craft. In one instance, he described encountering a human-like figure dressed in a blue uniform, who communicated with him telepathically and led him through a series of rooms before returning him to Earth.

The Travis Walton abduction case has been the subject of numerous investigations by UFO researchers and skeptics alike. Many aspects of the case lend credibility to Walton's story, including the consistency of the accounts provided by Walton and the other witnesses, as well as the results of the polygraph tests. Additionally, Walton's physical condition upon his return - including weight loss and dehydration - supports the idea that he experienced a genuine abduction event.

Over the years, Walton's story has been scrutinized and debated, yet it remains one of the most compelling and well-documented abduction cases in history. The Walton abduction case offers valuable insights into the potential existence of ancient aliens and their interactions with humans. The incident raises important questions about the nature of these beings, their motivations, and the technology they may possess. As such, the Travis Walton abduction remains an essential case study for proponents of the ancient alien theory.

Hybridization accounts form a fascinating and controversial subset of alien abduction experiences. Many individuals who report encounters with extraterrestrial beings also claim to have been subjected to reproductive experiments, resulting in the creation of hybrid offspring. These accounts suggest that extraterrestrial beings may be interested in the genetic manipulation and combination of human and alien DNA, potentially as part of a larger plan to create a hybrid species with enhanced capabilities or to ensure the survival of their own species. This section will explore these accounts in greater detail, examining their implications for the existence of ancient aliens and their potential motivations for engaging in such experiments.

Many individuals who recount their abduction experiences describe undergoing invasive reproductive procedures at the hands of extraterrestrial beings. These procedures often involve the collection of genetic material, such as sperm or eggs, through methods that may be both physically and emotionally traumatic for the abductee. In some cases, abductees report being implanted with embryos or having embryos removed from their bodies.

Proponents of the ancient alien theory argue that the consistency and specificity of these accounts, as well as the prevalence of such experiences across different cultures and time periods, may suggest that extraterrestrial beings have been engaging in reproductive experiments with humans for millennia. This could potentially lend credence to the idea that ancient alien encounters may have resulted in the creation of hybrid offspring in the past, just as they are purportedly doing in the present.

In addition to describing the reproductive procedures they have undergone, some abductees also report being taken aboard spacecraft to meet their hybrid children. These hybrid beings are often described as possessing a mix of human and alien features, such as large eyes, elongated skulls, and telepathic abilities. In some cases, abductees report forming emotional connections with their hybrid offspring and express concern for their wellbeing.

The existence of such hybrid beings, if true, could provide further evidence for the ancient alien theory, as it would suggest that extraterrestrial beings have been successful in creating viable hybrid offspring through their genetic experiments with humans. This would also raise intriguing questions about the motivations behind these hybridization efforts and the potential roles that these hybrid beings may play in the larger context of extraterrestrial interactions with humanity.

The accounts of alien hybridization have significant implications for the ancient alien theory. If extraterrestrial beings are indeed engaging in reproductive experiments with humans and creating hybrid offspring, this could suggest that ancient alien encounters may have involved similar activities. Such a possibility would lend further support to the idea that ancient myths and legends of gods mating with humans and producing hybrid offspring might be rooted in genuine encounters with extraterrestrial beings.

Furthermore, the hybridization accounts raise important questions about the motivations of extraterrestrial beings for engaging in these experiments. Are they attempting to create a hybrid species with enhanced capabilities, to ensure the survival of their own species, or for some other purpose entirely? As our understanding of genetics and the potential for human-alien hybridization advances, future research in this area may help to shed light on these questions and provide valuable insights into the nature and objectives of ancient alien interactions with humanity.

Accounts of alien contact and hybridization have long fascinated scientists and psychologists, sparking ongoing debate and inquiry. While some researchers approach these accounts with skepticism, others maintain that these experiences hold valuable insights into the potential existence of ancient aliens and their interactions with humanity. By examining the various scientific and psychological perspectives on alien contact and hybridization, we can better understand the potential evidence and implications for the existence of ancient extraterrestrials.

From a biological standpoint, the possibility of hybridization between humans and extraterrestrial beings raises intriguing questions about the nature of life in the universe. If ancient aliens indeed visited Earth and interbred with humans, this could suggest that life on other planets might share common genetic building blocks, such as DNA or RNA. This commonality would enable the creation of viable hybrid offspring, supporting the idea of a shared ancestry or convergent evolution among extraterrestrial species.

Additionally, the reported physical evidence of alien contact, such as implants and bodily marks, may be seen as an indication of the advanced biotechnology possessed by ancient aliens. These alleged artifacts, when analyzed, sometimes reveal materials and properties not found in nature, further fueling speculation about their extraterrestrial origins.

Anthropologists have long studied the myths and legends of ancient civilizations, seeking to understand the cultural and societal implications of these stories. The recurring theme of divine beings descending from the heavens to interact with humans, found in numerous cultures worldwide, has led some researchers to propose that these myths may have been inspired by actual encounters with ancient aliens.

In this view, ancient astronauts may have played a crucial role in shaping human culture, knowledge, and technological advancement. These extraterrestrial beings may have been perceived as gods, and their interactions with humans recorded in the myths and legends passed down through generations. By analyzing these stories from an anthropological perspective, we can uncover potential evidence for ancient alien contact and its impact on human civilization.

Astronomy and the search for extraterrestrial life are intrinsically linked, as the discovery of habitable exoplanets and potential alien signals could provide evidence for the existence of intelligent life beyond Earth. Some researchers argue that accounts of alien contact and hybridization, if true, could offer valuable insights into the nature of extraterrestrial civilizations and their potential interest in Earth.

For example, the star map reported by Betty and Barney Hill during their alleged abduction experience has been analyzed and proposed to represent the Zeta Reticuli star system, which is known to contain at least one potentially habitable exoplanet. This information may offer clues about the origins of ancient aliens and their motivations for visiting Earth, as well as the likelihood of encountering extraterrestrial life in our cosmic neighborhood.

Advancements in quantum physics and multidisciplinary approaches to the study of consciousness have opened new avenues for understanding the nature of reality and the potential existence of other dimensions or parallel universes. Some researchers propose that alien contact and hybridization experiences might involve interactions with beings from these alternate realities, rather than traditional extraterrestrial life.

The quantum perspective proposes that these beings could possess advanced knowledge of the fundamental nature of reality and the ability to manipulate space-time, which might account for the seemingly supernatural aspects of some alien encounters. By exploring the potential links between quantum physics, consciousness, and alien contact, we can gain a more comprehensive understanding of the possibilities for ancient extraterrestrial interactions with humanity.

The various scientific and psychological perspectives on alien contact and hybridization offer valuable insights into the potential existence of ancient aliens and their interactions with humanity. By examining these accounts through the lenses of biology, anthropology, astronomy, and quantum physics, we can better understand the evidence and implications for the existence of ancient extraterrestrials.

As research in these fields continues to progress, it is crucial to remain open to new evidence and interdisciplinary approaches to the study of alien contact and hybridization. While skepticism is a vital aspect of scientific inquiry, so too is the willingness to explore the unknown and consider unconventional ideas. By doing so, we can continue to uncover potential evidence for ancient alien encounters and deepen our understanding of the universe and our place within it.

Future research may focus on the development of advanced technologies that could detect extraterrestrial signals, decipher ancient artifacts or texts, and further explore the nature of reality and consciousness. These advancements could provide crucial insights into the existence of ancient aliens and their potential impact on human history and culture.

In the meantime, accounts of alien contact and hybridization serve as a reminder of the vastness of the universe and the enduring human fascination with the unknown. As we continue to seek answers to the age-old questions surrounding our origins and our place in the cosmos, the possibility of ancient alien contact and hybridization remains an intriguing and thought-provoking area of study. By embracing both skepticism and curiosity, we can continue to explore this enigmatic phenomenon and its potential implications for our understanding of the universe and ourselves.

The extraterrestrial hypothesis proposes that some or all reported alien encounters may represent genuine contact with extraterrestrial beings. This hypothesis has gained traction among researchers and enthusiasts alike due to the consistency and global nature of such reports, as well as the presence of physical evidence in some cases. Advocates of this hypothesis argue that these experiences cannot be easily dismissed as fabrications or delusions, and instead provide valuable clues about ancient alien contact and the potential existence of hybrid species.

One of the key arguments in favor of the extraterrestrial hypothesis is the remarkable consistency observed in alien encounter reports. Individuals from diverse cultural backgrounds and locations have described strikingly similar experiences, often involving abduction, medical examinations, and communication with non-human entities. This consistency lends credibility to the notion that these encounters represent genuine contact with extraterrestrial beings rather than mere fabrications or psychological phenomena.

Alien encounters have been reported in nearly every corner of the globe, transcending geographical and cultural boundaries. This widespread distribution of encounter reports suggests a phenomenon that is not limited to specific regions or cultural contexts, further supporting the extraterrestrial hypothesis. If these encounters were purely psychological or cultural constructs, one would expect a more localized distribution of reports, with experiences varying significantly between different cultures and regions.

In some cases, physical evidence has been reported in association with alien encounters. This evidence includes unexplained scars or marks on abductees' bodies, the presence of strange substances or implants, and even trace evidence at alleged landing sites. The existence of such physical evidence challenges the notion that these encounters are merely fabrications or psychological events and bolsters the argument for genuine extraterrestrial contact.

Many alien encounter reports involve descriptions of advanced technology and knowledge that far surpasses current human understanding. Abductees have reported witnessing advanced propulsion systems, medical devices, and other technologies that defy conventional explanation. Additionally, some encounters involve the imparting of information or knowledge to the experiencer, often concerning subjects such as astronomy, physics, or environmental issues. This aspect of alien encounters lends further support to the extraterrestrial hypothesis, as it suggests contact with beings possessing knowledge and technology beyond our current capabilities.

The extraterrestrial hypothesis not only offers a potential explanation for modern alien encounters but also has significant implications for the possibility of ancient alien contact and hybridization. If extraterrestrial beings have been visiting Earth and interacting with humans throughout history, as suggested by the consistency and global nature of alien encounter reports, it is conceivable that ancient alien contact may have occurred as well. This raises intriguing questions about the role of extraterrestrial beings in shaping human history, culture, and even our genetic makeup, through potential hybridization events.

The extraterrestrial hypothesis provides a compelling argument in favor of genuine contact with extraterrestrial beings, both in the modern era and throughout human history. The consistency, global distribution, and physical evidence associated with alien encounters, along with the presence of advanced technology and knowledge, lend credibility to this hypothesis and challenge the notion that these experiences are mere fabrications or psychological events. As researchers continue to investigate the phenomenon of alien encounters, the extraterrestrial hypothesis will likely remain an essential area of inquiry, offering insights into the potential existence of ancient aliens and their influence on our world.

The psychosocial hypothesis proposes that accounts of alien contact and hybridization are the result of psychological and cultural factors, rather than genuine encounters with extraterrestrial beings. While this perspective offers a plausible explanation for some aspects of alien encounter reports, there are several limitations to the psychosocial hypothesis that may favor the existence of ancient aliens.

Sleep paralysis, a phenomenon where an individual is temporarily unable to move or speak upon waking or falling asleep, is often accompanied by vivid hallucinations. These hallucinations, known as hypnagogic (when falling asleep) and hypnopompic (when waking up) hallucinations, can sometimes involve otherworldly entities or abduction scenarios. However, this explanation falls short in accounting for the consistency of alien encounter narratives across various cultures and throughout history.

Moreover, not all reported alien encounters take place during sleep or in states of altered consciousness, with many occurring while the individuals are fully awake and conscious. This suggests that sleep-related phenomena cannot fully explain the wide range of alien contact and hybridization accounts.

The psychosocial hypothesis suggests that alien encounter accounts may be influenced by the absorption of cultural narratives about extraterrestrial life. While it is true that popular culture has undoubtedly shaped our collective understanding of aliens, this explanation does not account for the striking similarities in alien encounter stories that predate modern media depictions of extraterrestrial beings.

Furthermore, some aspects of alien contact and hybridization accounts are remarkably consistent across cultures and time periods, despite significant differences in cultural context and exposure to external influences. This consistency may suggest a common origin for these accounts, potentially pointing towards actual encounters with ancient aliens.

One of the most significant limitations of the psychosocial hypothesis is its inability to account for the physical evidence sometimes associated with alien contact and hybridization experiences. Cases involving unexplained scars, implants, or genetic anomalies are difficult to reconcile with a purely psychological or cultural explanation.

Additionally, the psychosocial hypothesis does not fully explain the level of detail and consistency present in many accounts of alien encounters. While some variation in stories would be expected if they were purely the product of human imagination, the level of consistency in these accounts suggests that they may be based on a shared reality, potentially rooted in genuine encounters with ancient aliens.

While the psychosocial hypothesis offers a compelling explanation for some aspects of alien encounter reports, it is limited in its ability to account for the full range of experiences, the consistency of narratives, and the physical evidence sometimes associated with these encounters. These limitations may favor the existence of ancient aliens and suggest that at least some alien contact and hybridization accounts are based on actual interactions with extraterrestrial beings.

The interdimensional hypothesis presents an alternative perspective on alien encounters, suggesting that these experiences may not involve beings from other planets in our universe, but rather entities from other dimensions or parallel universes. This hypothesis proposes that these interdimensional beings have advanced abilities, such as manipulating space-time and reality, which could explain the often inexplicable and seemingly supernatural aspects of many alien encounter accounts. While this hypothesis remains speculative, it has gained traction among some researchers and ancient alien theorists, offering intriguing possibilities for understanding these enigmatic experiences.

The concept of parallel universes, or the multiverse, has its roots in quantum mechanics, a branch of physics that deals with the behavior of matter and energy on the smallest scales. One interpretation of quantum mechanics, known as the Many Worlds Interpretation, suggests that every possible outcome of a quantum event actually occurs, with each outcome realized in a separate, non-communicating parallel universe. This idea provides a theoretical foundation for the interdimensional hypothesis, positing that countless parallel universes, each with their own unique properties and life forms, may exist alongside our own.

Throughout human history, there have been numerous accounts of encounters with beings that seem to defy the laws of our universe. These entities, often described as gods, angels, or other supernatural beings, have been reported to possess extraordinary abilities, such as shape-shifting, telepathy, and control over natural elements. Some ancient alien theorists argue that these accounts may, in fact, be descriptions of encounters with interdimensional beings, who, due to their advanced capabilities, were perceived as divine or otherworldly by our ancestors.

The interdimensional hypothesis also raises the possibility of the existence of portals or gateways between dimensions or parallel universes. Ancient texts and legends often describe mysterious doorways, gateways, or other passages that lead to other realms or worlds. Some researchers have suggested that these accounts may represent actual portals used by interdimensional beings to travel between universes, allowing them to interact with humans and potentially influence our history and culture.

As our understanding of the universe and our technological capabilities continue to advance, we may one day develop the means to communicate with or even travel to other dimensions or parallel universes. Recent developments in theoretical physics, such as string theory and M-theory, have already begun to explore the possibility of higher dimensions and the potential for interdimensional communication or travel. If these advancements were to come to fruition, they could provide direct evidence of the existence of interdimensional beings and offer new insights into the nature of ancient alien encounters.

The interdimensional hypothesis offers a compelling alternative perspective on the nature of alien encounters, suggesting that these experiences may involve contact with beings from other dimensions or parallel universes rather than extraterrestrial life from within our own universe. While this hypothesis remains speculative, it presents intriguing possibilities for understanding the seemingly supernatural aspects of many alien encounter accounts and may, in time, shed new light on our understanding of the potential existence of ancient aliens and their influence on our world.

A compelling aspect of the ancient alien theory is the presence of physical evidence associated with alien encounters and hybridization. This evidence, which includes unexplained bodily marks, unusual implants, and unique genetic material, lends credibility to the notion that extraterrestrial beings may have visited and interacted with humanity throughout history.

Many individuals who claim to have experienced alien encounters, particularly abductions, report discovering unexplained scars or marks on their bodies following these events. In some cases, these marks appear overnight, and the individuals have no recollection of how they were incurred. Researchers and proponents of the ancient alien theory argue that these scars may be evidence of extraterrestrial experimentation or attempts at hybridization.

In addition to unexplained scars and marks, some abductees report the presence of strange substances or foreign objects within their bodies. These implants, often small and metallic, have been recovered from various locations within the abductees' bodies. When analyzed, these implants have been found to consist of unusual materials or exhibit extraordinary properties, such as being resistant to rust, emitting electromagnetic fields, or having isotopic ratios inconsistent with terrestrial materials. This evidence has led some researchers to suggest that these implants may be of extraterrestrial origin, possibly serving as tracking devices or tools for genetic manipulation.

Cases of individuals possessing unique genetic traits or anomalies have also been presented as potential evidence of ancient alien hybridization. These individuals, who often claim to be the product of alien-human hybridization, have been found to possess genetic traits that deviate significantly from the human norm. In some instances, these traits include unusual blood types, atypical skull structures, or even the presence of non-human DNA sequences.

One notable example is the Starchild Skull, discovered in Mexico in the 1930s. This skull exhibits several atypical cranial features, including an enlarged brain cavity and a unique bone composition. Furthermore, DNA analysis has revealed the presence of non-human DNA within the skull, suggesting that it may represent the remains of a hybrid being. While the origins and nature of the Starchild Skull remain controversial, it serves as a fascinating example of the potential physical evidence for ancient alien hybridization.

Beyond the evidence related directly to individuals, researchers have also pointed to various artifacts and structures as potential physical evidence of ancient alien contact and hybridization. These objects, which often exhibit advanced technology or construction techniques that seem out of place for their historical context, have been suggested as evidence of extraterrestrial influence or the work of hybrid beings.

Examples of such artifacts and structures include the Antikythera Mechanism, an ancient Greek device believed to be an astronomical calculator, and the massive stone structures of Pumapunku in Bolivia, which exhibit precise stonework and advanced engineering that challenge conventional explanations of their construction.

The presence of physical evidence associated with alien encounters and hybridization lends credibility to the ancient alien theory. While skeptics may argue that alternative explanations exist for these phenomena, the possibility of extraterrestrial interaction with humanity remains an intriguing and plausible explanation. As we continue to study these cases and uncover new evidence, our understanding of the potential existence of ancient aliens and their influence on human history may grow, further strengthening the case for their presence and interaction with our species throughout history.

The presence of unexplained implants and bodily marks in individuals claiming to have been abducted by aliens is a fascinating aspect of contact phenomena that has intrigued researchers and fueled speculation about the existence of ancient aliens. These physical anomalies, which often cannot be easily explained by conventional medical or scientific knowledge, may represent compelling evidence of extraterrestrial experimentation or attempts at hybridization.

Several alleged alien implants have been recovered from abductees and subjected to scientific analysis. In some cases, these implants have been found to consist of unusual materials, such as meteoric iron or rare isotopes, which are not commonly found in objects of terrestrial origin. Additionally, some implants have exhibited strange properties, such as the ability to emit radio frequencies, resistance to corrosion, or an inexplicable ability to move within the host's body.

The presence of these implants, combined with their unusual composition and properties, has led some researchers to argue that they may be of extraterrestrial origin. This hypothesis is supported by the fact that many abductees report being subjected to medical examinations and procedures during their encounters, suggesting that the implants may serve a specific purpose related to alien experimentation or hybridization efforts.

While the precise function of alien implants remains a matter of speculation, several theories have been proposed to explain their potential role in extraterrestrial contact and hybridization. Some researchers suggest that the implants may serve as tracking devices, allowing aliens to monitor the movements and activities of abductees. Others argue that the implants may be designed to gather physiological data, such as blood pressure, body temperature, or hormonal levels, to better understand human biology and facilitate the creation of hybrid offspring.

Another possibility is that the implants may serve to alter or manipulate the abductee's genetic material, resulting in the development of new abilities or traits that are characteristic of hybrid beings. This theory is supported by numerous accounts of abductees reporting changes in their physical or mental abilities following their encounters, including heightened psychic or telepathic powers, increased intelligence, or improved physical strength and agility.

In addition to the presence of implants, many abductees report discovering unexplained scars, marks, or bruises on their bodies following their encounters. These markings often appear in symmetrical patterns, suggesting a deliberate and intentional origin rather than random accidents or self-inflicted injuries.

Some researchers argue that these bodily marks may be the result of invasive medical procedures performed by extraterrestrial beings during abduction experiences. They contend that these markings may serve as additional evidence of extraterrestrial experimentation or hybridization efforts, further supporting the hypothesis that ancient aliens have interacted with humanity throughout history.

The presence of unexplained implants and bodily marks in individuals claiming to have been abducted by aliens offers tantalizing evidence in favor of the existence of ancient aliens and their ongoing efforts to experiment with or create hybrid beings. While skeptics may argue that these phenomena can be explained by psychological factors or cultural influences, the unusual composition and properties of these implants, combined with corroborating accounts of bodily marks and invasive procedures, suggest that these physical anomalies warrant further investigation as potential evidence of extraterrestrial contact and hybridization.

The existence of genetic anomalies in certain individuals and archaeological findings has been a source of fascination for those who support the Ancient Alien Theory. Proponents argue that these unusual genetic traits may be evidence of ancient alien hybridization or intervention. The Starchild Skull serves as one such example, but there are other instances that warrant exploration as well.

The Starchild Skull, discovered in Mexico in the 1930s, has been the subject of ongoing debate due to its atypical cranial features and the presence of non-human DNA. The skull, which belonged to a child, exhibits an unusually large cranial capacity, flattened rear skull, and shallow eye sockets. Some researchers have suggested that these features could be the result of a genetic condition, while others argue that the skull's unique characteristics point to extraterrestrial origins.

Genetic testing of the Starchild Skull has revealed the presence of both human and non-human DNA. While contamination of the samples cannot be ruled out, the presence of non-human DNA has fueled speculation that the skull may represent evidence of ancient alien hybridization. Further research and testing are needed to determine the true origins of the Starchild Skull, but its unusual characteristics continue to intrigue researchers and proponents of the Ancient Alien Theory.

The Paracas Skulls, discovered in Peru, are another example of unusual cranial features that have led some researchers to suggest a connection to ancient alien hybridization. These elongated skulls, which date back to around 300 BCE, were initially thought to be the result of artificial cranial deformation practiced by the Paracas culture. However, some researchers argue that certain features of the skulls, such as their large cranial volume and unusual suture patterns, cannot be explained by cranial deformation alone.

DNA testing of the Paracas Skulls has revealed genetic anomalies that are not found in typical human populations. Some researchers contend that these genetic differences may be the result of ancient alien intervention or hybridization. While more research is needed to fully understand the origins of the Paracas Skulls, their unusual features have generated significant interest in the possibility of ancient alien contact.

Certain blood types, such as the Rh-negative blood type, have been proposed as potential evidence for ancient alien hybridization. The Rh-negative blood type is relatively rare, occurring in only around 15% of the global population. It is characterized by the absence of the Rh antigen, a protein found on the surface of red blood cells in most people.

The origins of the Rh-negative blood type remain a mystery, with some researchers suggesting that it may have resulted from a genetic mutation. However, proponents of the Ancient Alien Theory argue that the Rh-negative blood type could be evidence of extraterrestrial intervention or hybridization. They point to the fact that Rh-negative individuals often exhibit unusual physical traits and are more likely to report alien abduction experiences.

While more research is needed to fully understand the origins of these genetic anomalies, their presence offers intriguing possibilities for ancient alien contact and hybridization. The unusual characteristics of the Starchild Skull, Paracas Skulls, and Rh-negative blood type provide tantalizing clues that may one day help to shed light on the potential existence of ancient aliens and their influence on humanity.

The phenomenon of alien contact and hybridization not only raises questions about the existence of ancient aliens but also points to the possibility of their significant influence on humanity's development. While some accounts can be attributed to psychological or cultural factors, a growing body of compelling evidence suggests that these phenomena may indeed represent genuine encounters with extraterrestrial beings.

Further research into these phenomena holds the potential to revolutionize our understanding of extraterrestrial life and the role ancient aliens may have played in shaping human history, culture, and technological advancements. By examining the accounts of alien contact and hybridization, along with the corresponding physical evidence, we can begin to piece together a more comprehensive picture of the possible interactions between ancient aliens and humans.

The existence of ancient aliens could have far-reaching implications for our understanding of human history and culture. If these beings did interact with our ancestors, they may have imparted knowledge, technology, and even genetic material, resulting in the rapid development of ancient civilizations and the emergence of hybrid species.

This could explain the sudden appearance of advanced technology and monumental architecture in ancient societies, such as the Egyptians, Sumerians, and Mayans. Further research into these connections could shed light on the true origins of human civilization and help us to better understand our shared cosmic heritage.

The search for ancient alien contact and hybridization can also drive advancements in modern technology and space exploration. As we continue to explore our solar system and beyond, we may uncover further evidence of past extraterrestrial interactions, or even discover current traces of alien life.

The development of advanced propulsion systems, communication technologies, and astrobiological research could all benefit from the study of alleged ancient alien encounters and their potential impact on our world. This knowledge may be crucial in helping us to establish future relationships with extraterrestrial species, should they exist.

A multidisciplinary approach to researching ancient alien contact and hybridization is essential for gaining a comprehensive understanding of these phenomena. By incorporating expertise from fields such as archaeology, anthropology, genetics, astronomy, and psychology, we can more effectively examine the evidence and evaluate its potential implications.

In conclusion, the enduring accounts of alien encounters and hybridization throughout human history, spanning from ancient legends to modern-day abduction experiences, suggest a possible link between extraterrestrial beings and the development of our species. The compelling evidence presented by some cases challenges conventional explanations and invites further investigation into the potential existence of ancient aliens.

By embracing a multidisciplinary approach and remaining open to the possibility of ancient alien contact and hybridization, we can not only uncover new insights into human history and culture but also drive advancements in technology and space exploration. As we continue to probe the mysteries of our past and explore the vast cosmos, the search for ancient aliens and their influence on our world remains a captivating and essential area of inquiry, inspiring both researchers and the general public alike in our quest to understand our place in the universe.

Chapter 6

Ancient Artifacts and Anomalies

Evidence of Extra-Terrestrial Intervention

The Ancient Alien Theory proposes that extraterrestrial beings have visited Earth throughout history, interacting with ancient civilizations and even influencing the course of human development. One of the most compelling arguments in favor of this theory is the existence of various ancient artifacts and architectural anomalies, which seem to display advanced technology and knowledge beyond what we believe ancient civilizations were capable of. This chapter will explore these artifacts and anomalies, presenting the evidence that supports the idea of extraterrestrial intervention in our past.

The Great Pyramid of Giza, also known as the Pyramid of Khufu, stands as a testament to the ingenuity and architectural prowess of the ancient Egyptians. However, proponents of the Ancient Alien Theory contend that the remarkable precision and advanced mathematical knowledge incorporated into the pyramid's design suggest the involvement of extraterrestrial beings.

The construction of the Great Pyramid required immense logistical and engineering expertise. Transporting the massive limestone and granite blocks, some of which weighed up to 80 tons, from quarries several miles away was a daunting task. Moreover, the ancient Egyptians would have needed to cut, move, and position these massive stones with remarkable precision, using only rudimentary tools such as ropes, wooden sleds, and copper chisels.

Ancient Alien Theory proponents argue that the techniques required to complete these tasks would have been beyond the capabilities of the ancient Egyptians. They suggest that extraterrestrial beings may have provided advanced technologies, such as levitation or anti-gravity devices, to facilitate the transportation and placement of the massive stones.

The Great Pyramid's dimensions encode sophisticated mathematical knowledge, which some believe is evidence of extraterrestrial involvement. For example, the ratio of the pyramid's base perimeter to its height is approximately equal to the value of 2π, indicating an understanding of the mathematical constant pi. Additionally, the pyramid's dimensions exhibit the golden ratio, a mathematical proportion found in various aspects of nature and art.

The pyramid's alignment to the cardinal points is also remarkably precise, with its sides aligned to within 0.15 degrees of true north, south, east, and west. This level of accuracy suggests advanced knowledge of astronomy and geodesy, which some argue would have been beyond the capabilities of the ancient Egyptians.

The Great Pyramid's internal structure adds further intrigue to the Ancient Alien Theory. The pyramid contains three main chambers: the King's Chamber, the Queen's Chamber, and the Subterranean Chamber, all of which are connected by a series of narrow passageways. The King's Chamber houses a granite sarcophagus, which is too large to have been brought into the chamber through the passageways, suggesting that it was placed there before the chamber was sealed.

The Grand Gallery, a steeply inclined passageway leading to the King's Chamber, has a unique corbelled design that has baffled engineers and architects for centuries. The purpose of this enigmatic structure remains unknown, fueling speculation that it may have served a purpose beyond mere aesthetics or burial customs.

Proponents of the Ancient Alien Theory argue that the Great Pyramid's intricate internal structure may have had a functional purpose, such as harnessing and focusing energy or serving as a communication device for extraterrestrial visitors. They suggest that the advanced knowledge required to design and construct these features indicates the involvement of beings with superior intelligence and technology.

The Great Pyramid of Giza's extraordinary precision, mathematical and astronomical knowledge, and enigmatic internal structure provide compelling evidence in favor of the Ancient Alien Theory. While skeptics maintain that these features can be explained by the remarkable capabilities of the ancient Egyptians, proponents of the theory argue that extraterrestrial beings may have played a crucial role in the construction of this awe-inspiring monument.

Puma Punku, a mysterious site situated in the highlands of Bolivia, has long been a source of fascination for historians, archaeologists, and proponents of the Ancient Alien Theory. This enigmatic complex, part of the larger Tiahuanaco site, is believed to have been constructed around 536 CE. What sets Puma Punku apart from other ancient sites is the extraordinary precision and craftsmanship exhibited in its stonework.

The stone blocks used at Puma Punku are massive, with some weighing up to 100 tons. The majority of these blocks are made from red sandstone and andesite, materials quarried miles away from the site. Transporting these colossal stones through the rugged Andean terrain would have been a monumental task in itself, even more so considering the limited technology available to the Tiwanaku civilization.

The stones at Puma Punku have been intricately carved with straight lines and sharp angles, displaying an astonishing level of precision. The complexity of the interlocking stone blocks suggests a high degree of planning and engineering prowess. Some blocks even feature a series of perfectly drilled holes, with consistent diameters and spacing, which would have been difficult to achieve using the tools and techniques known to be available to the Tiwanaku people.

Proponents of the Ancient Alien Theory argue that the advanced construction techniques displayed at Puma Punku are evidence of extraterrestrial involvement. The precision and complexity of the stone cuts suggest the use of advanced technology, such as lasers or high-frequency sound waves, which would have been beyond the capabilities of the Tiwanaku people.

Furthermore, the intricate system of interlocking blocks implies a level of engineering expertise that seems out of place for the time period. Some researchers have suggested that this knowledge may have been imparted to the Tiwanaku civilization by extraterrestrial beings, who may have assisted in the planning and construction of Puma Punku.

One of the most intriguing features of Puma Punku is the presence of numerous H-shaped blocks. These blocks, which are found throughout the site, are precisely cut and display a uniformity that is difficult to explain. The function of these H-shaped blocks remains a mystery, leading some Ancient Alien Theory proponents to speculate that they may have served a purpose related to extraterrestrial technology or communication.

The Enigma of Puma Punku's Construction and Demise

Adding to the mystery surrounding Puma Punku is the fact that the site appears to have been abandoned before its completion. The ruins of the site are strewn with partially carved blocks and unfinished structures, suggesting that the construction process was halted abruptly. Ancient Alien Theory proponents speculate that this sudden abandonment may be linked to the departure of extraterrestrial beings who were involved in the construction process.

The advanced engineering techniques and enigmatic features of Puma Punku serve as a compelling argument in favor of the Ancient Alien Theory. The precision and complexity of the stone cuts, the intricate interlocking system, and the mysterious H-shaped blocks all point to the possibility of extraterrestrial involvement. As we continue to explore the depths of human history, Puma Punku will undoubtedly remain a significant piece of evidence in the ongoing debate surrounding the existence of ancient aliens.

The Antikythera Mechanism, a remarkable ancient artifact discovered in a shipwreck off the coast of Greece in 1901, has mystified researchers and historians for over a century. This ancient analog computer, dating back to around 150-100 BCE, showcases an extraordinary level of sophistication and precision, leaving many to wonder how such an advanced device could have been developed during the ancient Greek era.

Composed of more than 30 bronze gears, the Antikythera Mechanism was used to predict astronomical events such as eclipses and the positions of celestial bodies. Its components display a level of miniaturization and complexity comparable to that of modern-day clocks, which is truly remarkable considering the time period in which it was created.

The sophistication of the Antikythera Mechanism raises questions about the origins of the knowledge and technology required to create it. Ancient Alien Theory proponents argue that the device serves as evidence of extraterrestrial intervention in human history. They contend that the expertise needed to develop such a complex mechanism was beyond the capabilities of ancient Greek engineers, suggesting that guidance or inspiration from extraterrestrial beings may have played a role in its creation.

Further supporting this theory is the fact that the Antikythera Mechanism is not an isolated example of advanced knowledge in the ancient world. Similar advanced astronomical knowledge can be found in various ancient civilizations, such as the Maya, Babylonians, and ancient Egyptians, who were all able to predict astronomical events with remarkable accuracy. This widespread presence of advanced astronomical knowledge in the ancient world lends credence to the idea that extraterrestrial beings may have shared their expertise with multiple civilizations across the globe.

Additionally, the Antikythera Mechanism's complex gear system indicates a deep understanding of gear ratios, which was not widely known or utilized in other technologies during that time. This knowledge of gear ratios and their application in creating precise, intricate machinery is another factor that supports the idea of extraterrestrial involvement.

Some proponents of the Ancient Alien Theory have also pointed out that the Antikythera Mechanism's purpose—predicting celestial events and tracking celestial bodies—could be evidence of a fascination with or reverence for extraterrestrial beings. The ancients may have developed the device to better understand the cosmos and the celestial beings that they believed influenced their lives and civilization.

The Antikythera Mechanism stands as a testament to the remarkable ingenuity and knowledge of the ancient Greeks. However, its complexity, sophistication, and purpose have led many Ancient Alien Theory proponents to argue that it serves as evidence of extraterrestrial intervention in human history. While skeptics may attribute the device's creation to the resourcefulness and intellect of the ancient Greeks, the advanced technology and knowledge displayed in the Antikythera Mechanism provide a compelling argument for the involvement of otherworldly beings in shaping the course of human history.

The Baghdad Battery, discovered in modern-day Iraq in the 1930s, is a fascinating artifact that has captivated both archaeologists and proponents of the Ancient Alien Theory. This curious object dates back to the Parthian period (250 BCE-224 CE) and comprises a clay jar, a copper cylinder, and an iron rod. When assembled, these components resemble a rudimentary galvanic cell or battery, suggesting that the ancient Parthians may have possessed a basic understanding of electricity.

The potential use of the Baghdad Battery for electroplating is particularly intriguing. Electroplating is the process of depositing a thin layer of one metal onto another using an electric current, and it requires a clear comprehension of the principles of electricity. Some researchers believe that the Parthians may have used these rudimentary batteries to electroplate gold onto silver objects, an advanced technique for that time.

Proponents of the Ancient Alien Theory argue that the existence of the Baghdad Battery is evidence of extraterrestrial intervention, as the knowledge of electricity and its applications would have been far beyond the capabilities of the ancient Parthians. It is essential to consider the historical context: during the Parthian period, the known world had not yet developed a formal understanding of electricity or its potential uses.

The Ancient Alien Theory suggests that extraterrestrial beings may have provided the necessary knowledge and technology to develop these primitive batteries. These beings, with their advanced understanding of the universe and its physical laws, could have imparted this knowledge to the Parthians, allowing them to create the Baghdad Battery and use it for electroplating.

One potential scenario for the transmission of electrical knowledge from extraterrestrial beings to the Parthians is through direct contact and interaction. It is possible that these advanced beings visited Earth during the Parthian period, interacting with the inhabitants and sharing their knowledge of electricity and its applications. This exchange of information could have occurred through physical demonstrations, oral teachings, or even telepathic communication.

Another possibility is that the Parthians discovered the knowledge of electricity through artifacts left behind by extraterrestrial visitors. These artifacts could have contained instructions, diagrams, or other forms of information that guided the Parthians in the creation of the Baghdad Battery. Alternatively, the Parthians may have reverse-engineered these artifacts, learning about electricity and its applications through careful observation and experimentation.

The Baghdad Battery represents a compelling piece of evidence in support of the Ancient Alien Theory. Its existence raises questions about how the Parthians, living in a time when the formal understanding of electricity had not yet been developed, could have created such a device. Proponents of the Ancient Alien Theory argue that extraterrestrial intervention provides a plausible explanation, suggesting that these advanced beings may have shared their knowledge of electricity with the ancient Parthians, enabling them to develop the Baghdad Battery and potentially use it for electroplating.

The Nazca Lines, a collection of vast geoglyphs found in the Nazca Desert of southern Peru, have captivated researchers and enthusiasts for decades. Created between 500 BCE and 500 CE, these enigmatic designs include depictions of animals, plants, and geometric shapes, some stretching over 1,200 feet in length. The Nazca people created the geoglyphs by removing reddish-brown iron oxide-coated pebbles from the surface, unveiling the light-colored earth beneath.

The purpose behind the Nazca Lines has long been debated, as they are only fully visible from the air. Proponents of the Ancient Alien Theory argue that these mysterious lines may have served as landing markers or communication signs for extraterrestrial visitors, as the lines' aerial visibility suggests the involvement of beings with advanced aerial technology.

Some of the Nazca Lines align with the positions of celestial bodies during significant astronomical events, such as solstices and equinoxes. This sophisticated understanding of astronomy further supports the idea that the Nazca people may have been in contact with extraterrestrial beings possessing advanced knowledge of the cosmos. The lines' celestial alignments could have been used to indicate specific times for extraterrestrial visitations or served as navigational aids for otherworldly spacecraft.

The sheer scale and precision of the Nazca Lines are remarkable, considering the lack of advanced technology available to the ancient Nazca civilization. Creating these massive geoglyphs would have required an extraordinary level of planning, coordination, and skill. The fact that these intricate designs were accurately executed without the aid of aerial observation suggests that the Nazca people may have received guidance from extraterrestrial beings with superior technological capabilities.

The flat, open terrain of the Nazca Desert would have been ideal for landing spacecraft, further supporting the Ancient Alien Theory. The geoglyphs' strategic placement in such a location could indicate that these lines served as markings for extraterrestrial visitors to land and interact with the Nazca people. The Nazca Lines might have been a means for the ancient civilization to maintain ongoing contact with otherworldly beings, fostering cultural and technological exchanges.

The Nazca culture may have created these geoglyphs as part of a religious or ritualistic practice related to their interactions with extraterrestrial beings. The Nazca Lines might have held sacred significance, representing the powerful beings who visited from the sky and bestowed their knowledge and technology upon the civilization. Alternatively, the lines could have been created to honor these beings or to request their return for further guidance and assistance.

The Nazca Lines' aerial visibility, advanced knowledge of astronomy, technological requirements for their creation, and the possibility of being landing sites for extraterrestrial spacecraft all support the Ancient Alien Theory. While the true purpose of the Nazca Lines remains a mystery, these fascinating geoglyphs continue to inspire researchers and enthusiasts to explore the possibility of ancient extraterrestrial involvement in human history.

Sacsayhuamán, an ancient Inca fortress located near Cusco, Peru, is a testament to the extraordinary architectural prowess of the Inca civilization. The most striking feature of this site is its massive stone walls, which display an unparalleled level of precision and skill in their construction. These walls not only captivate historians and archaeologists but also serve as compelling evidence for proponents of the Ancient Alien Theory, who argue that extraterrestrial beings may have provided the necessary knowledge and technology to build these impressive structures.

The stones used in the construction of Sacsayhuamán's walls are enormous, with some weighing over 100 tons. The task of quarrying these massive stones alone would have been a monumental challenge for the ancient Incas, who lacked the advanced tools and machinery available to modern-day builders. The quarries from which the stones were sourced are located miles away from the construction site, further complicating the transportation process. Ancient Alien Theory proponents argue that the successful quarrying and transportation of these colossal stones suggest the involvement of extraterrestrial beings with access to advanced technologies, such as anti-gravity devices or powerful energy sources capable of lifting and moving the stones with ease.

The stones in the Sacsayhuamán walls are not only massive but also intricately cut and fitted together with remarkable precision. The walls were constructed without mortar, relying instead on the perfect alignment of each stone to ensure stability. The stones are so tightly fitted together that not even a sheet of paper can be inserted between them. This level of precision is astonishing, especially when considering the size and irregular shapes of the stones.

The advanced construction techniques displayed at Sacsayhuamán are difficult to reconcile with the tools and knowledge available to the ancient Inca civilization. Traditional stone-cutting methods, such as hammering and chiseling, would have been insufficient to achieve the level of precision observed in the walls. Proponents of the Ancient Alien Theory argue that the construction of these walls would have required the use of advanced technologies, such as high-frequency sound waves or laser cutting tools, which would have been beyond the capabilities of the ancient Incas.

The construction of the Sacsayhuamán walls also presents a knowledge gap that is difficult to explain without considering the possibility of extraterrestrial intervention. The ancient Incas had no known system of writing or formal engineering education, making the acquisition and transfer of the knowledge required to design and build such impressive structures even more perplexing. Ancient Alien Theory proponents argue that this knowledge gap could be explained by the direct intervention of extraterrestrial beings, who may have imparted their advanced engineering knowledge to the Incas, allowing them to create these extraordinary structures.

The Sacsayhuamán walls stand as an enigma in the annals of human history, challenging our understanding of the capabilities of ancient civilizations. The size and weight of the stones, the precision with which they were cut and fitted together, and the knowledge gap that surrounds the construction process all point to the potential involvement of advanced, otherworldly beings in the creation of these remarkable structures. As we continue to explore our past and seek answers to the mysteries that surround us, the Ancient Alien Theory remains a compelling and thought-provoking force in our quest to understand our origins and our place in the cosmos.

The Piri Reis Map, created by the Ottoman admiral and cartographer Piri Reis in 1513, is a highly accurate representation of the world at that time. The map is particularly fascinating because it appears to show the coastline of Antarctica, a continent that was not officially discovered until 1820. Furthermore, the map depicts Antarctica as ice-free, which suggests that it was created based on information from a time when the continent was not covered in ice—a period dating back millions of years.

The map also demonstrates a level of precision in its depiction of the coastline of South America and the relative positions of various landmasses that seems to surpass the cartographic knowledge and capabilities of the early 16th century. Additionally, the map appears to incorporate information from a wide range of sources, including ancient maps that have long been lost to history.

Some proponents of Ancient Alien Theory argue that the accuracy and advanced knowledge displayed in the Piri Reis Map could only have been obtained with the help of extraterrestrial beings, who may have provided the cartographer with detailed geographical information that was not available to humans at that time. They assert that the inclusion of an ice-free Antarctica in the map suggests that the extraterrestrial beings had access to information from a distant past, long before any known human civilization could have documented it.

Another intriguing aspect of the Piri Reis Map is its use of the Mercator projection, a cartographic technique that was not developed until nearly 70 years after the creation of the map. This projection allows for the accurate representation of landmasses on a flat surface, maintaining relative sizes and shapes, and is considered a significant advancement in the field of cartography. The presence of this projection in the Piri Reis Map raises questions about how such advanced techniques could have been employed at a time when they were not yet known to the cartographic community.

Furthermore, some researchers have pointed out that the Piri Reis Map appears to depict the Earth as seen from a high altitude, suggesting that the map's creators had access to aerial views of the planet. This perspective would have been unattainable for humans in the early 16th century, further supporting the idea that extraterrestrial beings may have played a role in the creation of the map.

The Piri Reis Map presents a compelling case for the involvement of ancient aliens in human history. The advanced knowledge and techniques exhibited in the map, such as the inclusion of an ice-free Antarctica, the use of the Mercator projection, and the possible aerial perspective, suggest that the cartographer may have had access to information and technology beyond the capabilities of his contemporaries. As a result, the Piri Reis Map serves as an intriguing piece of evidence in support of the Ancient Alien Theory, fueling the ongoing debate surrounding the existence of extraterrestrial beings and their potential interactions with ancient human civilizations.

The Stone Spheres of Costa Rica, also known as Las Bolas, are a remarkable collection of over 300 nearly perfect spherical stones scattered throughout the Diquis Delta region. Ranging in size from a few centimeters to over 2 meters in diameter, these enigmatic spheres are believed to have been created by the indigenous people of the area between 600 CE and 1000 CE.

The precise method used to create these perfectly round stones remains a mystery, as the indigenous people who inhabited the region during this time period are not known to have had access to the advanced tools or technology required to craft such perfect spheres. Moreover, the stones are made from a hard material called granodiorite, which would have been difficult to shape using primitive tools.

The purpose of the Stone Spheres of Costa Rica is still unknown, as they do not appear to have any clear functional or symbolic use. Some spheres have been found arranged in lines or geometric patterns, while others were discovered in isolation. This has led to a variety of theories regarding their purpose, ranging from astronomical alignments to social status symbols.

Some proponents of Ancient Alien Theory propose that the Stone Spheres of Costa Rica are evidence of extraterrestrial intervention. They argue that the advanced knowledge and technology required to create such perfect spheres must have been provided by beings from another world. These otherworldly visitors may have shared their knowledge and tools with the indigenous people, allowing them to create these remarkable artifacts.

In support of this theory, some point to the fact that similar spherical stones have been found in other parts of the world, such as the stone spheres in Bosnia, which are also shrouded in mystery. This global distribution of seemingly unrelated yet similar artifacts may suggest a common source of knowledge or influence, possibly originating from extraterrestrial visitors.

Additionally, the mysterious nature of the Stone Spheres of Costa Rica aligns with other unexplained ancient artifacts and structures, such as the Nazca Lines and Puma Punku, which are also cited as evidence of extraterrestrial involvement in human history. The presence of these spheres, combined with the lack of a clear explanation for their creation and purpose, continues to fuel speculation about the possible involvement of ancient aliens in the development of human civilizations.

While the origin and purpose of the Stone Spheres of Costa Rica remain a mystery, their existence provides yet another intriguing piece of evidence in support of the Ancient Alien Theory. By considering the advanced knowledge and technology that would have been required to create these perfect spheres, as well as their mysterious purpose and global distribution, it becomes increasingly plausible that extraterrestrial beings may have played a role in the development of ancient cultures around the world.

The crystal skulls are a collection of enigmatic artifacts made from clear or milky quartz, purportedly dating back thousands of years. These mysterious objects have captivated the imagination of researchers, historians, and proponents of the Ancient Alien Theory alike, as the technology required to create such intricate and precise objects seems to have been beyond the capabilities of ancient civilizations. In this section, we will delve deeper into the mystery surrounding the crystal skulls, exploring the evidence that suggests extraterrestrial involvement in their creation and the implications of this possibility for our understanding of human history.

The most famous of the crystal skulls is the Mitchell-Hedges skull, discovered in the 1920s in present-day Belize. This skull is remarkable for its life-like appearance, including a detachable lower jaw that allows the jaw to move like that of a real human skull. The Mitchell-Hedges skull is not the only crystal skull that has been found; numerous other skulls, varying in size and design, have been discovered in locations around the world, including Mexico, Central America, and Europe.

The crystal skulls are fascinating artifacts due to their incredible level of detail and precision. They are made from a single piece of quartz crystal, which is one of the hardest materials known to man, making the process of carving them extremely difficult. Furthermore, the skulls display a level of detail and accuracy that would be challenging to achieve even with modern technology, such as the precisely carved eye sockets, nasal cavities, and teeth.

The technology required to create the crystal skulls has long been a subject of debate among researchers. The hardness of quartz crystal would have made it extremely difficult to carve with the tools and techniques available to ancient civilizations. Additionally, the level of precision displayed in the skulls would have required a thorough understanding of human anatomy, as well as advanced knowledge of optics to achieve the lifelike appearance of the finished product.

Proponents of the Ancient Alien Theory argue that the complexity and sophistication of the crystal skulls point to extraterrestrial involvement in their creation. They suggest that the skulls may have been crafted using advanced technology, such as high-frequency sound waves or lasers, provided by extraterrestrial beings. This technology would have allowed the ancient artisans to carve the crystal with unparalleled precision and detail, creating artifacts that continue to astound researchers and historians to this day.

The purpose of the crystal skulls remains a mystery, with various theories suggesting that they were used for religious ceremonies, healing rituals, or as receptacles for the storage of knowledge. Ancient Alien Theory proponents propose that the crystal skulls may have been gifts from extraterrestrial visitors, serving as tangible evidence of their advanced technology and knowledge. Some even speculate that the skulls may contain encoded information or messages from these otherworldly beings, waiting to be unlocked by those who possess the necessary understanding and technology.

The crystal skulls, with their intricate details and advanced craftsmanship, continue to captivate and mystify those who study them. The technology required to create such objects seems to have been beyond the capabilities of ancient civilizations, lending credence to the idea that extraterrestrial beings may have played a role in their creation. As we continue to explore the depths of our past and the mysteries of the universe, the crystal skulls will undoubtedly remain a compelling piece of evidence in the ongoing debate surrounding the Ancient Alien Theory and our place in the cosmos.

The existence of ancient artifacts and architectural anomalies that display advanced technology and knowledge has fueled the debate surrounding the Ancient Alien Theory. While skeptics argue that these remarkable achievements can be explained by the ingenuity and resourcefulness of ancient civilizations, proponents of the theory maintain that extraterrestrial intervention may provide a more plausible explanation.

By examining the Great Pyramid of Giza, Puma Punku, the Antikythera Mechanism, the Baghdad Battery, the Nazca Lines, the Sacsayhuamán Walls, Stone Spheres of Costa Rica, The Piri Reis Map, and the crystal skulls, we have uncovered an array of compelling evidence that points towards the involvement of advanced, otherworldly beings in our past.

The Stone Spheres of Costa Rica, for example, are almost perfectly round, large stone balls dating back to 600 CE. These spheres, some weighing as much as 16 tons, display a level of precision in their shaping and placement that is hard to explain using only the tools and knowledge available to the ancient Diquís culture. Proponents of Ancient Alien Theory argue that these spheres could have been created with the help of extraterrestrial beings possessing advanced technology.

The Piri Reis Map, created in 1513 by the Ottoman admiral and cartographer Piri Reis, depicts the coastlines of South America, Africa, and Antarctica with remarkable accuracy, especially considering the limited knowledge of the time. Ancient Alien Theory proponents suggest that the map may have been based on information provided by extraterrestrial beings who had a comprehensive understanding of Earth's geography.

The combined evidence from these diverse sites and artifacts paints a picture of ancient civilizations that may have been in contact with extraterrestrial beings, benefiting from their advanced knowledge and technology. This perspective challenges conventional explanations and invites us to entertain the possibility that our ancestors were not alone in their endeavors.

As our understanding of the universe and human history continues to expand, the Ancient Alien Theory will undoubtedly remain a significant and thought-provoking force in our quest for answers about our origins and our place in the cosmos. By keeping an open mind and considering alternative explanations for the wonders of the ancient world, we may one day unlock the secrets of our past and gain a deeper understanding of our species' incredible journey through time.

Embracing the Ancient Alien Theory also has the potential to inspire future generations to explore the universe and seek out new knowledge. The possibility of extraterrestrial life visiting Earth in the past encourages us to look beyond our planet, fostering a sense of wonder and curiosity about the cosmos. This sense of wonder, coupled with a growing body of evidence supporting the existence of ancient aliens, can ultimately drive scientific advancements and our continued search for the truth about our place in the universe.

Chapter 7

The Role of Governments and Science

The Cover-up of Extraterrestrial Life

The possibility of extraterrestrial life has captivated humanity's imagination for centuries. While the Ancient Alien Theory proposes that advanced beings have visited Earth throughout history, some proponents of this theory argue that governments and scientific institutions have actively suppressed or dismissed evidence of extraterrestrial life to maintain the status quo. This chapter will explore the various instances that have fueled this belief, focusing on alleged government cover-ups, UFO sightings, and the role of scientific institutions in shaping public opinion about extraterrestrial life and its connection to ancient aliens.

The Roswell Incident, which took place in July 1947, remains one of the most controversial and debated UFO events in history. Proponents of the

Ancient Alien Theory argue that the evidence surrounding the incident points to a cover-up of extraterrestrial involvement. This expanded section delves deeper into the events, witness accounts, and subsequent investigations related to the Roswell Incident.

On July 4, 1947, a rancher named Mac Brazel discovered a large amount of strange debris scattered across his property near Roswell, New Mexico. The debris included metallic fragments, rubber-like material, and thin, lightweight beams inscribed with indecipherable symbols. Brazel reported his discovery to the local sheriff, who then notified the nearby Roswell Army Air Field.

On July 8, the Roswell Daily Record published a front-page story with the headline "RAAF Captures Flying Saucer On Ranch in Roswell Region." The article featured a statement from the Roswell Army Air Field claiming that they had recovered a "flying disc." However, the very next day, the military retracted the statement and claimed that the wreckage was merely a downed weather balloon.

Numerous witnesses, both military and civilian, have come forward over the years with accounts of the Roswell Incident that contradict the official weather balloon explanation. Some military personnel claimed to have seen the debris and described it as unlike anything they had ever encountered. The material was said to be extremely lightweight, yet incredibly strong, and impervious to heat or damage. Furthermore, the debris reportedly possessed unusual properties, such as the ability to return to its original shape after being crumpled or bent.

Several witnesses also claimed to have seen alien bodies at the crash site or during transport to a secret facility. Descriptions of the bodies typically included small, humanoid figures with large heads and almond-shaped eyes. These accounts have been largely dismissed by skeptics as the product of rumors and hoaxes, but Ancient Alien Theory proponents argue that they provide further evidence of extraterrestrial involvement in the Roswell Incident.

The U.S. government's rapid response to the Roswell crash, along with their conflicting statements and explanations, has fueled suspicions of a cover-up. It is believed that the military quickly secured the crash site, removed the debris and any alien bodies, and transported them to a secret facility, such as Area 51 or Wright-Patterson Air Force Base, for further study and analysis.

Critics of the weather balloon explanation argue that the military's actions were inconsistent with the recovery of a mundane object like a weather balloon. They contend that the level of secrecy and urgency displayed by the military suggests that the object was of far greater significance and likely of extraterrestrial origin.

The Roswell Incident has become a cornerstone event for proponents of the Ancient Alien Theory, as it offers potential evidence of extraterrestrial visitation in modern times. If the Roswell crash did involve an extraterrestrial craft, it would lend credence to the idea that advanced beings have been visiting Earth throughout history, possibly influencing ancient civilizations and shaping human development.

By examining the Roswell Incident through the lens of the Ancient Alien Theory, researchers can draw connections between modern UFO sightings and historical accounts of encounters with otherworldly beings. This perspective encourages a broader understanding of humanity's interactions with extraterrestrial life and the potential impact of these encounters on our past, present, and future.

The Roswell Incident remains a key event in the debate surrounding the existence of ancient aliens and the possibility of a government cover-up. The inconsistencies in the official explanation, along with the numerous witness accounts and the military's secretive actions, have provided fertile ground for speculation and investigation. The incident serves as a prime example of how alleged encounters with extraterrestrial life can challenge our understanding of human history and prompt us to reconsider the possibility of advanced beings visiting Earth.

As new information continues to emerge, and as technological advancements enable us to explore the cosmos further, the Roswell Incident will likely remain a significant event in the ongoing discussion of ancient aliens and their potential influence on our world. By maintaining an open-minded approach to these questions, we can foster a greater understanding of the mysteries that surround our existence and encourage a spirit of curiosity and inquiry into the unknown.

Ultimately, the Roswell Incident serves as a reminder that the search for answers about our origins, our place in the universe, and the potential existence of extraterrestrial life is a complex and ever-evolving pursuit. Whether or not the Roswell crash was indeed an encounter with ancient aliens, it stands as a testament to the power of human curiosity and our unrelenting desire to explore the unknown.

Project Blue Book was an official United States Air Force investigation into UFO sightings that took place from 1952 to 1969. The project's stated purpose was to determine if UFOs posed a threat to national security. However, many proponents of the Ancient Alien Theory believe that the real aim of Project Blue Book was to discredit the phenomenon and suppress public interest in extraterrestrial life. In this section, we will delve deeper into the project's history, methodology, and findings, examining how they may have contributed to an environment that favored the dismissal of evidence supporting the existence of ancient aliens.

Project Blue Book was the third and final iteration of the United States Air Force's investigations into UFO sightings. It followed Project Sign (1947-1949) and Project Grudge (1949-1952), both of which laid the groundwork for the government's skeptical approach to UFO phenomena. Project Blue Book was led by a series of officers, most notably Captain Edward J. Ruppelt, who coined the term "Unidentified Flying Object" (UFO) and attempted to bring a more scientific and open-minded approach to the investigations.

However, Ruppelt's departure from the project in 1953 marked a shift towards greater skepticism and dismissal of UFO sightings. Subsequent leadership, such as that of Major Hector Quintanilla, focused on finding conventional explanations for sightings rather than considering the possibility of extraterrestrial life.

Project Blue Book employed a range of scientific and investigative methods to analyze UFO sightings. This included interviewing witnesses, examining photographic and video evidence, and consulting with experts in various fields, such as astronomy, meteorology, and aviation. Despite these efforts, the project's methodology has been criticized for its bias towards finding conventional explanations for UFO sightings, even in cases where such explanations were tenuous or unconvincing.

Out of the 12,618 UFO sightings investigated by Project Blue Book, 701 remained unexplained. The project's final report, known as the Condon Report, concluded that there was no evidence to support the existence of extraterrestrial life and that UFO sightings could be attributed to conventional explanations, such as misidentifications or natural phenomena.

Critics of Project Blue Book argue that the Air Force's dismissal of these unexplained cases indicates a lack of genuine interest in the possibility of extraterrestrial life. They contend that this bias towards conventional explanations served to maintain the status quo and suppress public interest in the existence of ancient aliens.

The closure of Project Blue Book in 1969 marked the end of the United States Air Force's official investigations into UFO sightings. However, the project's legacy continues to influence public perception of UFO phenomena and the possibility of extraterrestrial life.

Proponents of the Ancient Alien Theory argue that Project Blue Book's skepticism and focus on conventional explanations for UFO sightings have contributed to a climate of dismissal and suppression regarding evidence of ancient alien contact. They believe that the project's methodology and conclusions have had a lasting impact on the scientific community and government institutions, leading to a reluctance to consider alternative theories about humanity's history and interactions with extraterrestrial life.

By examining the history and findings of Project Blue Book, it becomes apparent that the project may have played a role in shaping public opinion and scientific inquiry in a manner that favored the dismissal of evidence supporting the existence of ancient aliens. As we continue to explore the depths of space and uncover the mysteries of our past, the legacy of Project Blue Book serves as a reminder of the importance of maintaining an open-minded approach to the possibility of extraterrestrial life and its potential influence on human history.

Given the criticisms of Project Blue Book's methodology and its potential bias towards conventional explanations, it is crucial to revisit the unexplained cases in a more open-minded and unbiased manner. By re-examining these cases, researchers may uncover evidence that could further support the Ancient Alien Theory or provide new insights into the nature of UFO phenomena.

With advancements in technology and scientific knowledge, it is possible that some of the previously unexplained cases could now be better understood. Furthermore, by applying a more comprehensive approach that considers alternative theories, such as the Ancient Alien Theory, researchers may be able to identify patterns or connections between cases that were previously dismissed or overlooked.

As the Ancient Alien Theory continues to gain traction and challenge conventional views of human history, it is essential to address the potential suppression of public interest in extraterrestrial life by government institutions and mainstream science. This can be achieved through promoting transparency, fostering open-minded inquiry, and encouraging interdisciplinary research into the possibility of ancient alien contact.

One way to address this suppression is by advocating for the release of previously classified or withheld information related to UFO sightings and extraterrestrial encounters. By making this information available to the public, researchers and proponents of the Ancient Alien Theory can analyze the data more thoroughly, potentially uncovering new evidence that supports the existence of ancient aliens.

Additionally, promoting open-minded dialogue between mainstream science and proponents of alternative theories, such as the Ancient Alien Theory, can help to break down barriers and foster an environment of intellectual curiosity and exploration. By encouraging interdisciplinary research and collaboration, we can work towards a more comprehensive understanding of humanity's place in the cosmos and the potential influence of extraterrestrial life on our history.

Project Blue Book's legacy of skepticism and dismissal of UFO sightings has had a lasting impact on public perception and scientific inquiry into the possibility of extraterrestrial life. By revisiting the project's unexplained cases and addressing the suppression of public interest in ancient aliens, we can promote a more open-minded approach to understanding the mysteries of our past and our potential interactions with extraterrestrial life. As we continue to explore the cosmos and uncover the secrets of our history, it is essential to maintain an open-minded perspective and be willing to consider alternative theories that challenge conventional views of human history and our place in the universe.

The Kecksburg UFO Incident remains one of the most intriguing and debated events in UFO history. Occurring on December 9, 1965, the incident began with a fireball streaking across the skies of several states in the U.S. and parts of Canada. Witnesses reported that the fireball changed direction several times before crashing into the woods near Kecksburg, Pennsylvania. This peculiar behavior, combined with the subsequent military response, has fueled speculation that the object was of extraterrestrial origin.

Numerous witnesses described the object as an acorn-shaped, metallic craft with strange markings or symbols on its surface, resembling hieroglyphics. Some witnesses even claimed to have seen the object partially embedded in the ground, with the surrounding trees and vegetation damaged from the impact. The consistent accounts of the object's appearance and the damage it caused to the environment lend credibility to the assertion that it was not a typical meteor, but rather an artificial, possibly extraterrestrial, craft.

The military's swift and extensive response to the Keckburg incident has raised suspicions of a cover-up. Within hours of the crash, armed military personnel arrived at the scene, cordoning off the area and restricting access to both civilians and local authorities. Witnesses reported seeing the object removed under a tarp on a flatbed truck, accompanied by a military convoy. This level of secrecy and the expeditious removal of the object indicate that there may have been a concerted effort to conceal evidence of extraterrestrial life.

The official explanation given by the U.S. government and Project Blue Book investigators was that the object was a meteor or space debris. However, this explanation fails to address the many inconsistencies in witness accounts and the peculiar behavior of the object as it traversed the sky. The dismissal of these discrepancies has only fueled speculation that the government is concealing the true nature of the incident.

The Kecksburg UFO Incident has been cited by proponents of the Ancient Alien Theory as potential evidence of extraterrestrial life and its involvement in human history. The strange markings on the object, described as hieroglyphic-like symbols, could suggest a connection to ancient civilizations that have been theorized to have had contact with extraterrestrial beings. Additionally, the acorn-shaped craft is reminiscent of depictions of ancient flying machines, such as the Indian Vimanas, further supporting the idea that extraterrestrial technology has influenced humanity throughout history.

The Kecksburg UFO Incident remains a compelling example of a potential government cover-up aimed at concealing evidence of extraterrestrial life. The object's unique appearance, the consistent witness accounts, and the military's extensive response all point to a phenomenon that cannot be easily dismissed as a meteor or space debris. As proponents of the Ancient Alien Theory continue to examine such incidents, they reveal the possibility of a deeper connection between humanity and extraterrestrial civilizations, challenging our understanding of our place in the cosmos.

In December 1980, a series of UFO sightings occurred in Rendlesham Forest, near two U.S. Air Force bases in the United Kingdom, RAF Woodbridge and RAF Bentwaters. This event, which has come to be known as the Rendlesham Forest Incident, is one of the most well-documented and compelling UFO encounters in history.

Over the course of three nights, numerous military personnel reported seeing a strange craft with bright lights and unusual markings in the forest. On the first night, two U.S. Air Force security patrolmen spotted unusual lights descending into the forest. They sought permission to investigate and encountered a small, triangular craft with a metallic appearance. The craft had white, blue, and red lights, and appeared to hover silently just above the ground. As they approached, the craft quickly maneuvered through the trees and disappeared.

On the second night, additional sightings were reported, and the deputy base commander, Lieutenant Colonel Charles Halt, led a team to investigate. Halt and his team observed strange lights in the sky and a pulsating red light that seemed to emit beams down to the ground. Halt recorded his observations on an audio tape, which has since become a crucial piece of evidence in the case.

On the third night, a serviceman named Jim Penniston claimed to have encountered the craft up close. He described it as black and triangular, with unusual symbols similar to Egyptian hieroglyphs. Penniston reported that he touched the craft and experienced a sudden download of binary code into his mind, which he later transcribed into a notebook. Years later, this binary code was analyzed and found to contain a message with coordinates pointing to various ancient sites around the world, further fueling speculation about a connection to ancient aliens.

In addition to the numerous eyewitness accounts, the Rendlesham Forest Incident is notable for the physical evidence left behind. The landing site was examined and found to have three indentations in the ground, forming an equilateral triangle. These indentations were consistent with the description of the craft provided by the witnesses. Moreover, the trees in the area showed signs of damage, and Geiger counter readings indicated increased radiation levels at the site.

Despite the compelling witness accounts and physical evidence, the U.S. and UK governments dismissed the incident as a misidentification of a nearby lighthouse, which was visible from the forest. Critics argue that this explanation is unsatisfactory and fails to account for the detailed descriptions of the craft, its movements, and the physical evidence left behind. The apparent reluctance of the governments to conduct a thorough investigation and the classification of related documents have led some to believe that the Rendlesham Forest Incident was an encounter with an extraterrestrial craft and that efforts were made to suppress this information.

The Rendlesham Forest Incident is often cited by proponents of the Ancient Alien Theory as evidence of ongoing extraterrestrial visitation. The binary code message allegedly received by Jim Penniston, with coordinates pointing to ancient sites around the world, suggests a possible connection between the craft and ancient alien visitors. These coordinates include locations such as the Great Pyramid of Giza, the Nazca Lines, and the ancient Sumerian city of Ur. This connection implies that the beings behind the Rendlesham Forest Incident may have been the same as those who interacted with ancient civilizations, providing further support for the Ancient Alien Theory.

The Battle of Los Angeles, also known as the Great Los Angeles Air Raid, is a compelling event in the history of UFO encounters and provides significant support for the Ancient Alien Theory. Occurring on the night of February 24-25, 1942, this incident involved the military's aggressive response to an unidentified object in the skies above Los Angeles. Despite the official explanation attributing the event to a weather balloon or "war nerves," the circumstances surrounding the Battle of Los Angeles suggest a possible extraterrestrial presence.

During the incident, military personnel and civilians alike reported seeing a large, slow-moving, and possibly metallic object in the sky. Descriptions of the object varied, with some witnesses claiming it was an oval or round craft, while others reported seeing a formation of smaller objects. In response to the sighting, the military launched a barrage of anti-aircraft artillery shells at the object for more than an hour, with some reports suggesting that over 1,400 rounds were fired.

The intensity of the military's reaction to the unidentified object raises questions about the official explanation. The sheer number of rounds fired and the extensive duration of the engagement suggest that the object was perceived as a genuine threat, rather than a simple misidentification of a weather balloon or a case of heightened war anxiety.

The U.S. government's explanations for the incident have been inconsistent and, in many cases, contradictory. Initially, the military attributed the event to a false alarm caused by an unidentified aircraft, possibly a commercial plane. Later, the official explanation shifted to a weather balloon, and then to "war nerves," suggesting that the military and civilian witnesses had simply imagined the entire encounter.

These inconsistent explanations do not adequately address the numerous witness accounts, the photographic evidence of the event, and the extensive military response. Many proponents of the Ancient Alien Theory argue that these discrepancies indicate a deliberate attempt to suppress evidence of extraterrestrial life and to maintain the status quo.

The Battle of Los Angeles is a prime example of an incident that supports the existence of ancient aliens or advanced extraterrestrial civilizations. The object's appearance and behavior, as well as the military's aggressive response, suggest that it was not a conventional aircraft or weather phenomenon. Furthermore, the government's apparent efforts to downplay the event and dismiss the numerous witness accounts imply a potential cover-up to conceal evidence of extraterrestrial life.

In the context of the Ancient Alien Theory, the Battle of Los Angeles may represent an encounter with an advanced civilization, either visiting or monitoring Earth for unknown reasons. The incident serves as a reminder of the potential presence of extraterrestrial life in our history and the ongoing efforts to suppress this information.

The Battle of Los Angeles remains a significant and intriguing event that supports the Ancient Alien Theory. The numerous witness accounts, the military's response, and the inconsistencies in the official explanations all point to the possibility of an extraterrestrial presence during this incident. By examining such events, we can continue to challenge conventional explanations and explore the potential influence of ancient aliens on our world.

The Phoenix Lights incident on March 13, 1997, remains one of the most significant and widely-discussed UFO sightings in modern history. Thousands of witnesses across Arizona reported seeing a massive V-shaped craft adorned with multiple lights silently gliding across the sky. The sheer number of witnesses, including the then-governor of Arizona, Fife Symington, lends credibility to the event, which has become an important case for proponents of the Ancient Alien Theory.

Not only did the Phoenix Lights incident involve a large number of witnesses, but the descriptions of the craft were remarkably consistent. People reported a huge, V-shaped object, often described as larger than a football field, that moved silently and at a low altitude. Many witnesses claimed that the object passed directly over their homes, allowing them to observe it closely.

The U.S. military's explanation that the lights were merely flares from training exercises has been widely criticized, as it does not account for the silent, slow-moving nature of the craft or its immense size. Furthermore, the military's response to the incident was delayed, with the flare explanation only being provided weeks after the event. This has led many to question the veracity of the official explanation and suspect a cover-up designed to suppress public interest in extraterrestrial life.

Proponents of the Ancient Alien Theory assert that the Phoenix Lights incident is an example of the ongoing presence of extraterrestrial beings on Earth. They argue that the craft observed during this event may be linked to the same advanced civilizations that visited our planet in ancient times and influenced the development of human society. The V-shaped craft could be a modern manifestation of the ancient Vimanas described in Hindu mythology or other advanced aerial vehicles mentioned in ancient texts.

Moreover, the Phoenix Lights incident occurred in an area known for its rich Native American history and ancient petroglyphs. Ancient Alien Theory proponents suggest that the sighting may have taken place near a location of historical significance to these ancient visitors, possibly indicating an ongoing connection between the extraterrestrial beings and our planet.

The government's dismissal of the Phoenix Lights as flares from military training exercises raises questions about their commitment to transparency and openness regarding the UFO phenomenon. Proponents of the Ancient Alien Theory argue that the government may have information about the existence of extraterrestrial life, ancient or otherwise, which they are actively trying to suppress. The Phoenix Lights incident serves as a prime example of the potential evidence for ancient aliens and their continued presence in our world.

The Phoenix Lights incident remains an important and widely-discussed event in the realm of UFO sightings and Ancient Alien Theory. The numerous witnesses, consistent descriptions of the V-shaped craft, and the inadequacy of the official explanation provide compelling reasons to consider the possibility of extraterrestrial involvement. As we continue to explore our past and strive to understand our place in the cosmos, events like the Phoenix Lights serve as a reminder of the potential connections between our world and the ancient visitors who may still be watching over us.

The Search for Extraterrestrial Intelligence (SETI) program has long been at the forefront of the quest to find intelligent life beyond Earth. By scanning the skies for radio signals, SETI aims to detect evidence of technologically advanced civilizations. However, proponents of the Ancient Alien Theory argue that this approach has inherent limitations that may hinder our understanding of extraterrestrial life and its potential influence on human history.

SETI's primary focus on radio signals is based on the assumption that extraterrestrial civilizations would use similar communication methods as humans. While this may seem logical, critics argue that this approach is anthropocentric and may overlook alternative forms of communication that advanced civilizations might employ. For instance, it is possible that extraterrestrial beings could utilize other parts of the electromagnetic spectrum or rely on more advanced technologies, such as quantum communication or telepathy, which would remain undetected by SETI's current methods.

By concentrating on the detection of radio signals, SETI's approach largely ignores other potential forms of evidence for extraterrestrial life, such as ancient artifacts, structures, and texts. Ancient Alien Theory proponents argue that these sources provide compelling indications of past interactions between humans and extraterrestrial beings. Examples include the precise construction of the Great Pyramid of Giza, the intricate stone carving techniques used at Puma Punku, and ancient texts describing advanced technologies and flying machines.

The dismissal of these alternative forms of evidence by mainstream scientific institutions, including SETI, has led some Ancient Alien Theory proponents to question whether there is an intentional effort to maintain a limited understanding of extraterrestrial life.

The scientific community often relies on established paradigms to interpret and understand new data. This adherence to established frameworks can sometimes lead to resistance against ideas that challenge prevailing views, such as the Ancient Alien Theory. Proponents of the theory argue that institutions like SETI may be influenced by a desire to protect existing scientific paradigms from disruption, and as a result, they may be less open to considering alternative forms of evidence.

This reluctance to explore alternative viewpoints may contribute to a limited understanding of extraterrestrial life, hindering our ability to recognize the potential influence of ancient aliens on human history.

In order to gain a more comprehensive understanding of extraterrestrial life and its potential impact on human history, it is essential for scientific institutions, including SETI, to adopt a more open-minded approach. By broadening the search parameters and considering alternative forms of evidence, such as ancient artifacts and texts, the possibility of discovering connections between ancient aliens and human civilization increases.

Furthermore, exploring new communication technologies and methods beyond radio signals could help scientists identify previously undetected forms of extraterrestrial communication, further expanding our understanding of the universe and our place within it.

While SETI has made significant contributions to the search for extraterrestrial life, proponents of the Ancient Alien Theory argue that its narrow focus on radio signals may overlook crucial evidence of past interactions between humans and extraterrestrial beings. By embracing a more open-minded approach and considering alternative forms of evidence, scientific institutions like SETI can potentially uncover new insights into the existence of ancient aliens and their influence on our world.

Mainstream scientific institutions, such as major universities and research organizations, have long been critical of alternative theories regarding extraterrestrial life, including the Ancient Alien Theory. While skepticism is an essential component of the scientific method, proponents of the Ancient Alien Theory argue that mainstream science's approach to alternative theories is overly dismissive and may ultimately hinder the pursuit of knowledge.

One reason that mainstream science often discredits alternative theories is the reluctance to challenge established paradigms. The prevailing scientific worldview is grounded in the belief that human civilization developed independently, without external intervention. As a result, the idea that advanced extraterrestrial beings may have influenced ancient cultures is seen as a direct challenge to this widely accepted perspective.

By dismissing the Ancient Alien Theory, mainstream scientific institutions may be inadvertently perpetuating a narrow understanding of human history that ignores potential evidence of extraterrestrial involvement. Critics argue that this reluctance to consider alternative explanations is a form of intellectual conservatism that stifles innovation and progress in the study of our past.

Another issue that arises in the relationship between mainstream science and alternative theories is the dismissal of anomalous evidence. Proponents of the Ancient Alien Theory often point to examples of advanced engineering in ancient structures, such as the precise alignment of the Great Pyramid of Giza or the intricate stonework of Puma Punku. These feats are difficult to explain using conventional archaeological and historical models, suggesting that the builders may have had access to advanced knowledge or technologies.

However, mainstream scientists often dismiss these anomalies as either coincidences or the result of unknown but human-based technologies. Critics argue that this approach is a form of confirmation bias, as it prioritizes explanations that fit within the established paradigm while discounting evidence that may challenge these assumptions.

One of the strengths of the Ancient Alien Theory is its interdisciplinary nature, drawing on fields such as archaeology, anthropology, astronomy, and even engineering to build a comprehensive picture of humanity's interactions with extraterrestrial beings. However, mainstream scientific institutions often operate within specialized silos, limiting collaboration and the sharing of ideas between disciplines.

This lack of interdisciplinary collaboration may contribute to the dismissal of alternative theories, as experts in one field may be unwilling or unable to consider evidence from other disciplines that could support the Ancient Alien Theory. By fostering a more collaborative environment, mainstream science could potentially be more open to considering alternative explanations and incorporating a broader range of evidence.

Finally, it's essential to consider the role of media and popular culture in shaping public opinion about alternative theories, such as the Ancient Alien Theory. Mainstream scientific institutions often benefit from the support of well-funded media outlets and respected publications, which amplify their perspectives and lend credibility to their views. On the other hand, alternative theories are often relegated to niche media outlets, leading to a perception that these ideas are less credible or fringe.

This imbalance in media representation can contribute to the dismissal of alternative theories by mainstream science, as the Ancient Alien Theory may be perceived as less legitimate due to its association with popular culture rather than rigorous academic inquiry. To overcome this barrier, proponents of the Ancient Alien Theory must continue to engage with the scientific community and present well-researched, evidence-based arguments that challenge the established paradigms.

The role of mainstream science in discrediting alternative theories like the Ancient Alien Theory is multifaceted, encompassing a range of factors such as established paradigms, the dismissal of anomalous evidence, lack of interdisciplinary collaboration, and media representation. By addressing these issues and fostering a more open-minded approach to the study of humanity's past, we may be able to gain a more comprehensive understanding of our history and the potential influence of extraterrestrial life on our civilization.

Engaging in dialogue with mainstream scientists and researchers can help promote open-mindedness and facilitate a more nuanced understanding of alternative theories. By presenting well-researched and evidence-based arguments, proponents of the Ancient Alien Theory can demonstrate the credibility of their claims and encourage others to consider the possibility of extraterrestrial influence on ancient cultures.

Fostering a spirit of interdisciplinary collaboration, proponents of the Ancient Alien Theory can break down the barriers between scientific disciplines and encourage a more holistic approach to the study of human history. This may lead to new insights and discoveries that could help support the theory and challenge conventional wisdom.

Publishing research in peer-reviewed journals, presenting at conferences, and participating in academic forums, proponents of the Ancient Alien Theory can help legitimize their ideas within the scientific community. This increased visibility may encourage more scientists to consider alternative theories and explore their implications for our understanding of human history.

Actively addressing misconceptions and misunderstandings surrounding the Ancient Alien Theory, proponents can help dispel the stigma associated with alternative ideas and promote a more balanced view of the evidence. This may involve debunking common myths, clarifying misconceptions, and engaging in respectful debate with critics.

Proponents of the Ancient Alien Theory can advocate for changes in scientific practices that would promote a more inclusive approach to the study of human history. This might include supporting funding for alternative research, encouraging interdisciplinary collaboration, and challenging the scientific community to be more open to unconventional ideas.

Taking these steps, proponents of the Ancient Alien Theory can work towards a more inclusive and open-minded scientific community that is willing to consider a broader range of evidence and ideas. In turn, this may lead to new discoveries and insights that challenge our understanding of human history and the potential influence of extraterrestrial life on our civilization.

The role of governments and scientific institutions in the alleged cover-up of extraterrestrial life remains a highly debated and complex subject. While concrete evidence of a large-scale conspiracy might not be available, a series of unexplained incidents, including the Roswell crash, the Rendlesham Forest Incident, and the Phoenix Lights, have continued to fuel suspicions that crucial information about extraterrestrial life is being deliberately withheld from the public.

Supporters of the Ancient Alien Theory argue that the skeptical attitude of mainstream science towards alternative theories of extraterrestrial life is indicative of intellectual gatekeeping and suppression of evidence. They contend that there is a wealth of information in ancient texts, artifacts, and structures that supports the idea of ancient alien visitations, but this evidence is often dismissed or overlooked due to prevailing scientific paradigms.

As we continue to explore the universe and our understanding of it grows, the debate surrounding the existence of ancient aliens and the alleged role of governments and scientific institutions in concealing their presence is unlikely to abate. To truly uncover the truth about our place in the cosmos and the potential influence of extraterrestrial life on our history, it is crucial to maintain an open-minded approach to these questions.

Embracing alternative theories, such as the Ancient Alien Theory, can foster a spirit of inquiry and encourage a more comprehensive understanding of humanity's origins. By challenging conventional wisdom and re-examining the evidence, we can potentially unlock new insights into our past, as well as our connection to the wider cosmos.

In the pursuit of truth, it is essential to remain vigilant and question the motives and actions of governments and scientific institutions that may have an interest in concealing information about extraterrestrial life. It is only through an unwavering commitment to openness, transparency, and critical examination that we can hope to unveil the truth about ancient aliens and their potential impact on human history.

Ultimately, the pursuit of knowledge about ancient aliens and their potential role in our history not only enriches our understanding of the past but also broadens our perspective on the universe we inhabit. By acknowledging the possibility of ancient extraterrestrial visitations and remaining open to alternative theories, we can foster a more inclusive and expansive view of our place in the cosmos and our connection to the mysteries that lie beyond our current understanding.

Chapter 8

The Search for Extraterrestrial Communication

SETI and Beyond

The search for extraterrestrial communication has been a topic of fascination and speculation for centuries. The possibility of otherworldly beings attempting to communicate with Earth through various means has fueled research into methods for detecting and deciphering potential signals from distant worlds. The connection to Ancient Alien Theory lies in the notion that evidence of past communication or even visitation from extraterrestrial beings could provide a foundation for modern-day search efforts.

The SETI Institute's mission to search for signs of extraterrestrial intelligence lends support to the Ancient Alien Theory, as it operates on the premise that advanced civilizations exist beyond our planet and may have attempted to communicate with Earth. By analyzing electromagnetic radiation from distant stars and galaxies, SETI researchers actively seek evidence that could reveal a historical or ongoing relationship between Earth and extraterrestrial beings.

SETI employs a variety of methodologies, including radio astronomy, optical SETI, and the study of exoplanets, in its quest to uncover potential signs of extraterrestrial communication. These methods are directly relevant to the Ancient Alien Theory, as they are aimed at detecting signals that could be indicative of extraterrestrial civilizations – civilizations that may have interacted with or influenced ancient human societies. Furthermore, the SETI Institute's commitment to exploring various avenues in the search for extraterrestrial communication highlights the possibility that ancient alien contact could have occurred through multiple channels, such as electromagnetic signals, laser pulses, or even physical visitations.

The SETI Institute collaborates with numerous institutions, researchers, and organizations worldwide to further the understanding of the cosmos and the potential existence of extraterrestrial life. This collaborative approach fosters the sharing of knowledge, resources, and expertise, creating a global network of scientists and enthusiasts committed to the pursuit of evidence for ancient alien contact. The widespread interest and involvement in SETI's mission demonstrate the significance and appeal of the Ancient Alien Theory to a broad audience and contribute to the ongoing search for answers about our place in the universe.

The SETI Institute's work has had a profound impact on popular culture and public perception of the possibility of extraterrestrial life. Through media coverage, documentaries, and public outreach programs, the Institute has helped to popularize the idea that intelligent life may exist beyond our planet and has shaped the way people view the cosmos. This influence extends to the Ancient Alien Theory, as the SETI Institute's work has fostered greater awareness and acceptance of the possibility of past or present contact with extraterrestrial civilizations. In this way, the SETI Institute plays a critical role in promoting the exploration of ancient alien contact and its potential implications for our understanding of human history and our place in the cosmos.

The Drake Equation's relevance to the Ancient Alien Theory lies in its attempt to quantify the probability of intelligent life existing elsewhere in the universe. By examining the various factors that contribute to the development and sustainability of extraterrestrial civilizations, the equation suggests that the existence of intelligent life outside of Earth is a distinct possibility. This supports the notion that ancient aliens may have visited or influenced human history in some way.

The rate of star formation in our galaxy has a significant impact on the likelihood of planets that could harbor life. With an estimated 200 to 400 billion stars in the Milky Way alone, the probability of planetary systems capable of supporting life increases. As the number of stars and planets increases, so too does the likelihood of ancient alien civilizations that could have interacted with Earth.

Another critical factor in the Drake Equation is the concept of habitable zones, or the regions around stars where conditions are just right for liquid water to exist on a planet's surface. The presence of liquid water is considered a prerequisite for life as we know it, and the discovery of exoplanets within these habitable zones strengthens the case for the existence of ancient alien civilizations. If other planets can provide the necessary conditions for life, the possibility of advanced beings capable of visiting or influencing Earth becomes more plausible.

The final components of the Drake Equation involve the probability of intelligent life evolving and becoming capable of interstellar communication. While the specific factors that contribute to the development of intelligent life remain unknown, the vastness of the universe and the sheer number of potential habitable planets suggest that it is statistically likely that other intelligent beings exist. If these civilizations developed advanced technologies, such as the capability for interstellar travel or communication, it is reasonable to assume that they may have made contact with Earth at some point in history.

The Drake Equation serves as a foundation for assessing the probability of intelligent civilizations existing in our galaxy, thereby supporting the Ancient Alien Theory. While the equation does not offer definitive proof of extraterrestrial life, it provides a framework for understanding the potential existence of advanced civilizations capable of interacting with Earth. The interconnectedness of factors such as star formation, habitable zones, and the development of intelligent life and communication technologies suggests that the possibility of ancient aliens is not only plausible but also worthy of further investigation. As our understanding of the universe expands, the relationship between the Drake Equation and the Ancient Alien Theory will continue to inform our quest for answers about the potential existence of extraterrestrial life and its impact on human history.

Radio telescopes such as the Arecibo Observatory in Puerto Rico and the Green Bank Telescope in West Virginia have been pivotal in the search for extraterrestrial communication. These massive instruments have facilitated the detection and analysis of radio signals from the depths of space, providing astronomers with the means to identify potential indicators of intelligent life. The importance of these radio telescopes is emphasized when considering the Ancient Alien Theory, as they offer a means to potentially uncover evidence of past or ongoing communication with extraterrestrial beings.

One of the most famous potential extraterrestrial signals ever detected is the "Wow!" signal, discovered in 1977 by astronomer Jerry R. Ehman using the Big Ear radio telescope at Ohio State University. This strong, narrowband radio signal lasted for 72 seconds and has never been observed again. The signal's unique characteristics and the inability to identify a natural source have led some proponents of Ancient Alien Theory to suggest that it could be evidence of extraterrestrial communication. Although no definitive link to alien civilizations has been established, the "Wow!" signal remains an intriguing mystery in the search for extraterrestrial intelligence.

Fast Radio Bursts (FRBs) are another example of enigmatic radio signals that have captured the attention of scientists and Ancient Alien Theory enthusiasts alike. FRBs are brief, high-energy radio signals that have been observed since 2007. These signals originate from distant galaxies, and their exact causes remain unknown. Some theories propose that FRBs could result from natural phenomena, such as neutron star collisions or magnetar flares. However, the possibility that these signals may be evidence of advanced extraterrestrial technology cannot be entirely ruled out. The ongoing study of FRBs and the search for their origins could potentially offer insights into the existence of intelligent alien life.

The ongoing development of radio astronomy and the construction of more advanced radio telescopes, such as the Square Kilometer Array (SKA) and the Next-Generation Very Large Array (ngVLA), promise to enhance our ability to detect and analyze radio signals from deep space. These cutting-edge instruments will allow researchers to probe even further into the cosmos and potentially uncover previously undetectable signals that may offer evidence of extraterrestrial communication. By refining our ability to identify and decipher radio signals, radio astronomy holds the potential to reveal previously undiscovered evidence supporting the Ancient Alien Theory.

Radio astronomy plays a crucial role in the search for extraterrestrial communication, offering the possibility of detecting signals that could indicate the presence of intelligent life beyond Earth. Instances such as the "Wow!" signal and Fast Radio Bursts, although not definitively linked to alien civilizations, lend credence to the possibility of extraterrestrial communication and the existence of ancient aliens. As radio astronomy continues to advance, the potential to uncover evidence supporting the Ancient Alien Theory grows, furthering our understanding of the universe and our place within it.

The concept of Optical SETI emerged from the realization that advanced extraterrestrial civilizations might be using more efficient and technologically sophisticated means of communication than radio waves. As laser technology has advanced rapidly on Earth, it is reasonable to assume that other civilizations might also develop and utilize laser-based communication systems. This line of thinking aligns with the Ancient Alien Theory, as it suggests that past contact with extraterrestrial beings could have involved advanced technologies such as laser communication, which would be consistent with the purported capabilities of ancient alien civilizations.

Laser communication offers several advantages over radio signals, which make it an attractive option for interstellar messaging. Laser beams can be focused into a tight, directed beam, allowing for more accurate targeting of distant star systems. This feature reduces the likelihood of signal loss due to dispersion and enables information to be transmitted more efficiently. Furthermore, laser signals are less prone to interference from cosmic background radiation and can carry a higher bandwidth, allowing for the transmission of more complex data. These attributes are consistent with the technological prowess that ancient aliens are presumed to have possessed, further supporting the possibility of past extraterrestrial communication via laser signals.

The detection of laser signals from extraterrestrial civilizations requires the use of specialized equipment and observational strategies. Optical telescopes equipped with high-speed photodetectors and narrow-band filters are used to scan the sky for pulsed or continuous laser signals. Due to the vastness of space and the numerous potential sources of laser signals, the search for extraterrestrial laser communication is both time-consuming and resource-intensive. However, the potential rewards of discovering definitive evidence of extraterrestrial laser communication would be significant, as it would provide strong support for the existence of ancient aliens and their advanced technological capabilities.

As technology continues to advance, the potential for detecting extraterrestrial laser signals is likely to improve. Future developments in telescope technology, photodetectors, and data analysis techniques may enhance our ability to detect and interpret potential laser signals from distant star systems. These advancements could ultimately provide the key to discovering evidence of extraterrestrial communication, lending further credence to the Ancient Alien Theory. By uncovering definitive proof of laser-based communication from other civilizations, we may be able to unlock the secrets of our own past and gain a deeper understanding of the potential interactions between ancient human civilizations and extraterrestrial beings.

The Fermi Paradox and the Great Silence present a compelling challenge to the Ancient Alien Theory, but several plausible explanations can reconcile this paradox while still supporting the idea of extraterrestrial contact in our past.

The vastness of the universe, with its billions of galaxies, each containing billions of stars and potentially even more planets, is a factor that may explain the Great Silence. The sheer scale of the cosmos means that detecting other civilizations becomes an incredibly daunting task. Moreover, the distances between stars are enormous, making the chances of detecting or intercepting communications from extraterrestrial beings significantly more difficult. Ancient Alien Theory proponents argue that past civilizations may have made contact with Earth, but due to the immense distances involved, they may have lost the ability to communicate or travel back to our planet.

Our current technology for detecting extraterrestrial signals is relatively limited, which may contribute to the Fermi Paradox. We primarily rely on radio and optical telescopes to search for extraterrestrial signals, but it is entirely possible that advanced civilizations utilize entirely different communication methods that we are not yet capable of detecting. Additionally, the window of detectability – the period during which a civilization's signals are detectable – might be relatively small compared to the vast timescale of cosmic history. Ancient Alien Theory suggests that the evidence for extraterrestrial contact in our past could be out there, but we have not yet developed the technology necessary to find it.

Another explanation for the Great Silence that supports the Ancient Alien Theory is the Zoo Hypothesis. This hypothesis proposes that advanced extraterrestrial civilizations are aware of humanity but have chosen not to interfere with our development, similar to the Prime Directive in the Star Trek universe. These civilizations may have made contact with Earth in the past, leaving behind traces of their presence, as proposed by Ancient Alien Theory. However, they may now be observing us from a distance, waiting for humanity to reach a certain level of technological or social maturity before reestablishing contact.

The Rare Earth Hypothesis suggests that the specific conditions necessary for the development of intelligent life are exceptionally rare, making Earth a special case in the cosmos. If this hypothesis is accurate, it would help explain the Great Silence. However, this does not necessarily contradict the Ancient Alien Theory. It could be that Earth is one of the few planets where intelligent life has developed, attracting the attention of extraterrestrial beings who visited and influenced our planet in the past. This would make Earth an outlier in the universe, but it would not negate the possibility of ancient alien contact.

The Fermi Paradox and the Great Silence present a significant challenge to the Ancient Alien Theory. However, considering the immensity of space and time, the limitations of our current technology, and alternative hypotheses like the Zoo Hypothesis and the Rare Earth Hypothesis, it is possible to reconcile the Fermi Paradox with the idea that Earth has been visited or influenced by advanced extraterrestrial civilizations in the past. As our understanding of the universe expands and our technology advances, we may eventually find definitive evidence that supports the Ancient Alien Theory and resolves the Fermi Paradox.

The Kardashev Scale provides a framework for understanding the potential capabilities of advanced extraterrestrial civilizations, which can lend support to the Ancient Alien Theory. By exploring the implications of each type of civilization on the scale, we can better understand how these advanced beings may have interacted with Earth in the past and left their mark on our history.

Type I civilizations, with their ability to harness and utilize the energy resources of their home planet, would possess a level of technological sophistication far beyond our current capabilities. Such civilizations could have visited Earth in the distant past, leaving behind traces of their advanced technology in the form of unexplained artifacts and structures. Proponents of the Ancient Alien Theory argue that the advanced engineering found in ancient megalithic structures, such as the pyramids of Egypt or the stone structures of Puma Punku, could be evidence of Type I extraterrestrial involvement in human history.

Type II civilizations would be capable of harnessing the energy of their host star, enabling them to engage in large-scale interstellar travel and colonization. With this level of technological prowess, these civilizations could have visited Earth in the past, influencing human development and shaping our history through direct interaction or indirect guidance. Ancient Alien Theory proponents point to ancient myths and legends of gods descending from the skies, as well as advanced knowledge of astronomy and mathematics found in various ancient cultures, as potential evidence of Type II extraterrestrial contact.

Type III civilizations, with the ability to harness the energy of an entire galaxy, would possess virtually unlimited resources and capabilities. Such advanced beings could have played a role in shaping the very fabric of our universe, creating conditions conducive to the development of intelligent life. This idea aligns with the Ancient Alien Theory, which suggests that extraterrestrial beings may have had a hand in the evolution of life on Earth, possibly through direct genetic intervention or by seeding our planet with the building blocks of life. The presence of complex DNA structures and the rapid emergence of intelligent life on Earth could be seen as evidence for the involvement of Type III civilizations in our history.

The Kardashev Scale provides a useful framework for understanding the potential capabilities of advanced extraterrestrial civilizations and their possible interactions with Earth throughout history. By considering the implications of each type of civilization on the scale, researchers can develop a more comprehensive picture of how these advanced beings may have influenced our past, lending support to the Ancient Alien Theory and inspiring further exploration of our cosmic origins.

The detection and study of exoplanets have been made possible through various advanced astronomical techniques, such as the transit method, radial velocity method, and direct imaging. These methods have led to the discovery of thousands of exoplanets, vastly expanding our understanding of the diversity of planetary systems in the universe. The multitude of exoplanets discovered, including those within the habitable zones of their host stars, has strengthened the Ancient Alien Theory by demonstrating the plausibility of other Earth-like worlds where intelligent life could have emerged and potentially visited Earth in ancient times.

The habitability of an exoplanet is determined by a combination of factors, including its size, atmospheric composition, distance from its host star, and the presence of liquid water. These factors are critical for the development and sustenance of life as we know it. The discovery of potentially habitable exoplanets, including those with Earth-like characteristics, bolsters the Ancient Alien Theory by providing more opportunities for life to have arisen elsewhere in the universe. The more habitable worlds we identify, the higher the likelihood that some of these planets may harbor intelligent life capable of interstellar travel and communication, further supporting the notion of ancient alien contact.

One of the key objectives of exoplanet research is the search for biosignatures, or signs of life, in the atmospheres of these distant worlds. By analyzing the atmospheric composition of exoplanets, astronomers can look for chemical imbalances that may indicate the presence of living organisms. The detection of biosignatures on exoplanets would provide compelling evidence for the existence of extraterrestrial life and lend further support to the Ancient Alien Theory. The discovery of such biosignatures would strengthen the argument that intelligent life has evolved elsewhere in the universe and may have contacted or visited Earth in ancient times.

The identification and study of potentially habitable exoplanets have significantly impacted the search for extraterrestrial communication. As we discover more worlds that could potentially support life, astronomers can prioritize these planets in their search for signals from extraterrestrial civilizations. By targeting potentially habitable exoplanets, researchers increase the likelihood of detecting communication from advanced civilizations. This targeted approach to the search for extraterrestrial communication lends further credence to the Ancient Alien Theory, as it increases the probability of discovering evidence that Earth has been visited or contacted by extraterrestrial beings in the past.

Future exoplanet missions, such as the James Webb Space Telescope and the European Space Agency's PLATO mission, will enable astronomers to further characterize the atmospheres and surfaces of potentially habitable worlds. These missions will provide more detailed information about the conditions on these exoplanets, allowing researchers to refine their assessments of habitability and the potential for life. As we continue to expand our knowledge of the universe and discover more habitable worlds, the case for the Ancient Alien Theory grows stronger, with an increasing number of potentially life-supporting planets pointing to the possibility of ancient extraterrestrial contact with Earth.

The concept of Active SETI is rooted in the idea that by actively reaching out to other civilizations, we might uncover evidence of ancient alien contact or even establish communication with extraterrestrial beings. Proponents of Active SETI argue that this approach is necessary in order to break the "Great Silence" that has thus far characterized our search for extraterrestrial intelligence. They posit that by sending out messages, we might elicit a response from civilizations that have been monitoring Earth or that have visited our planet in the past, thereby providing further support for the Ancient Alien Theory.

Throughout human history, civilizations have sought to establish contact with distant peoples, often leading to the exchange of knowledge, resources, and cultural practices. In this context, the idea of reaching out to extraterrestrial civilizations can be seen as a continuation of this innate human desire for exploration and connection. Proponents of Active SETI often cite historical examples of successful contact between disparate civilizations as evidence that the potential benefits of METI outweigh the risks. They argue that the discovery of ancient alien contact could lead to transformative advancements in science, technology, and our understanding of the universe.

While the potential consequences of broadcasting our presence to the cosmos are indeed a valid concern, supporters of Active SETI maintain that the benefits of establishing communication with extraterrestrial beings may outweigh the risks. They argue that the potential for collaboration and the exchange of knowledge with advanced civilizations could lead to unprecedented advancements in human society, possibly even uncovering evidence of past ancient alien contact. In this view, the possibility of discovering the truth about the Ancient Alien Theory justifies the risks associated with Active SETI.

In order to address the ethical concerns associated with Active SETI, some proponents suggest that an international consensus on the content and format of messages sent into space should be established. By involving a diverse range of stakeholders, including scientists, philosophers, and representatives from various cultural and religious backgrounds, a more balanced and inclusive approach to METI can be developed. This cooperative approach could help ensure that the potential benefits of establishing communication with extraterrestrial beings are maximized while minimizing the risks associated with broadcasting our presence to the cosmos.

Ultimately, the debate surrounding Active SETI reflects the broader questions and uncertainties that surround the Ancient Alien Theory. As researchers continue to explore the possibility of extraterrestrial communication, the potential implications of ancient alien contact will undoubtedly remain a topic of discussion and debate. By considering the ethical dimensions of our search for extraterrestrial intelligence and exploring the potential benefits of establishing communication with other civilizations, we may be better positioned to uncover the truth about our possible connections to the cosmos and the role that ancient aliens may have played in shaping human history.

The development of advanced propulsion systems, such as those based on antimatter or nuclear fusion, could revolutionize interstellar travel and open up new possibilities for the search for extraterrestrial communication. Antimatter propulsion systems, which involve the annihilation of matter and antimatter to release energy, could potentially enable spacecraft to achieve speeds close to the speed of light. Similarly, nuclear fusion propulsion, which involves harnessing the energy generated by the fusion of atomic nuclei, could offer a more efficient and sustainable means of space travel compared to traditional chemical rockets.

Another promising avenue for interstellar travel is the development of solar sail technology, as demonstrated by the Breakthrough Starshot initiative. This ambitious project aims to send a fleet of small, lightweight spacecraft, propelled by powerful laser beams, to the nearest star system, Alpha Centauri, at speeds of up to 20% the speed of light. By utilizing the momentum transfer from photons to propel the spacecraft, solar sails could offer a feasible method for traversing vast interstellar distances without the need for conventional fuel. If successful, projects like Breakthrough Starshot could pave the way for more efficient exploration of neighboring star systems and increase the likelihood of detecting evidence of ancient alien contact.

As we develop the capability to travel closer to other star systems, our chances of detecting and deciphering potential extraterrestrial signals will likely increase. The proximity to other star systems could enhance our ability to intercept electromagnetic radiation, such as radio or laser signals, emitted by advanced civilizations. This would allow for a more direct and efficient search for evidence of extraterrestrial life, potentially uncovering definitive proof of ancient alien contact.

Interstellar travel could also enable the direct search for physical evidence of ancient alien contact. By exploring other star systems and their planets, we may discover extraterrestrial artifacts or structures that provide tangible evidence of advanced civilizations that have interacted with our own. This direct search for extraterrestrial artifacts could lend further support to the Ancient Alien Theory and shed new light on the role that extraterrestrial beings may have played in the development of human civilization.

The future of interstellar travel and the search for extraterrestrial communication will likely require international collaboration and the pooling of resources, knowledge, and expertise from various fields. By working together, scientists, engineers, and space agencies from around the world can overcome the immense challenges associated with interstellar exploration and potentially unlock the secrets of our cosmic neighbors. This collaborative effort could ultimately provide the key to discovering definitive evidence of ancient alien contact, furthering our understanding of the role that extraterrestrial beings may have played in shaping our history and the development of human civilization.

In conclusion, the search for extraterrestrial communication has captivated the imaginations of scientists, researchers, and enthusiasts alike, driven by the desire to uncover the mysteries of the universe and to explore the potential existence of other intelligent life forms. This pursuit, incorporating various scientific disciplines and methodologies, serves as a testament to humanity's innate curiosity and unrelenting spirit of discovery.

When considering the Ancient Alien Theory, it is essential to approach the subject with open-minded inquiry and a willingness to entertain unconventional ideas. By doing so, we can better appreciate the possibility that our planet's history may be intertwined with extraterrestrial civilizations, who may have played a role in shaping our world and its development. The search for extraterrestrial communication, when viewed through the lens of the Ancient Alien Theory, presents an exciting opportunity to delve deeper into the complex tapestry of our past and to challenge conventional wisdom.

As our understanding of the cosmos deepens, we continue to make remarkable discoveries that reshape our perceptions of the universe and its inhabitants. From the detection of potentially habitable exoplanets to advancements in interstellar travel technology, these breakthroughs bring us closer to answering the age-old question of whether we are alone in the universe. In the context of the Ancient Alien Theory, these discoveries and advancements offer tantalizing clues that may eventually lead to definitive evidence of past or present communication with extraterrestrial beings.

The implications of the Ancient Alien Theory extend far beyond the scientific community, influencing our culture, philosophy, and understanding of human history. The notion that extraterrestrial civilizations may have visited or influenced Earth in the distant past challenges traditional narratives and encourages us to consider alternative explanations for ancient artifacts, structures, and myths. In this way, the Ancient Alien Theory serves as a catalyst for intellectual growth and fosters an environment in which unconventional ideas can be explored and debated.

As the search for extraterrestrial communication continues, the Ancient Alien Theory will undoubtedly remain an integral part of humanity's ongoing quest to understand our place in the universe and the potential existence of other intelligent life forms. By embracing the possibility of ancient alien contact and encouraging open-minded inquiry into the mysteries of our past, we can expand our understanding of the cosmos and its inhabitants, fostering a greater appreciation for the rich and complex history of our planet.

Chapter 9

Alien Abduction and Close Encounter Experiences

Psychological Effects and Impacts

Alien abduction and close encounter experiences are not limited to specific cultures or time periods, making them a truly universal phenomenon. Individuals from all corners of the globe and throughout history have reported encounters with otherworldly beings, suggesting a consistency that cannot be easily dismissed. This universality lends credibility to the idea that these experiences may be evidence of extraterrestrial contact, as opposed to mere fabrications or misinterpretations of natural phenomena.

Many individuals who have reported alien abduction or close encounter experiences describe not only vivid visual and auditory details but also physical and emotional sensations that accompany these encounters. The physical sensations often include feelings of paralysis, levitation, or being transported to another location, while emotional responses can range from extreme fear and panic to a sense of awe and wonder. The intensity and specificity of these sensations lend further weight to the argument that such experiences may indeed be evidence of contact with extraterrestrial beings.

One of the most striking aspects of alien abduction and close encounter experiences is the consistency in the descriptions of extraterrestrial beings. Across cultures and time periods, abductees have reported remarkably similar characteristics of these beings, including their physical appearance, behavior, and the nature of their interactions with humans. This consistency suggests that these encounters may be more than mere figments of the imagination, as it is unlikely that so many individuals from different backgrounds would independently create such similar descriptions.

Many abductees recall encountering advanced technology and alien environments during their experiences, often describing intricate and highly advanced devices or systems that are beyond the scope of human understanding. These reports often include details that are consistent with theoretical or cutting-edge technologies, such as anti-gravity propulsion, advanced medical procedures, or telepathic communication. The fact that these descriptions align with current scientific understanding and speculation lends credibility to the idea that these encounters may be genuine and indicative of extraterrestrial contact.

For some individuals, alien abduction or close encounter experiences serve as a catalyst for personal transformation, leading to a deeper understanding of themselves and the universe. Abductees may report experiencing heightened intuition, psychic abilities, or even spiritual awakening following their encounters with extraterrestrial beings. This transformative aspect of alien abduction experiences could be seen as evidence of the profound impact that contact with advanced extraterrestrial civilizations can have on human consciousness and personal development.

The Greys, the small, grey-skinned beings with large, almond-shaped eyes, are frequently reported by abductees from different cultures and geographical locations. The consistency of their descriptions across abduction accounts is seen by proponents of the Ancient Alien Theory as strong evidence for the existence of extraterrestrial beings. The Greys are often depicted as being highly intelligent and technologically advanced, possessing a deep understanding of human biology and the ability to communicate telepathically.

Another common theme in alien abduction accounts is the experience of missing time or unexplained gaps in memory. Abductees often report that they cannot remember the entire duration of their encounters or that they have difficulty recalling specific details of their experiences. The consistency of these memory gaps across abduction accounts suggests that they may be more than mere coincidences, and could be indicative of an intentional manipulation of human memory by extraterrestrial beings.

During their encounters, many abductees report experiencing sleep paralysis or other altered states of consciousness. These altered states can manifest as an inability to move or speak, vivid hallucinations, and a heightened awareness of one's surroundings. The prevalence of these experiences in abduction accounts is seen by some as evidence of the use of advanced technology or mind control techniques by extraterrestrial beings, which allow them to manipulate human perceptions and induce these altered states.

The intense emotional responses reported by abductees during their encounters, such as fear, fascination, and awe, are another common element in alien abduction and close encounter accounts. These emotions may be the result of the overwhelming nature of the experience, as well as the abductees' realization that they are in the presence of beings and technology far beyond their understanding. The consistency of these emotional responses across abduction accounts lends further support to the notion that these experiences may be genuine encounters with extraterrestrial beings.

The commonalities in abduction and close encounter accounts, such as the appearance of the Greys, missing time, altered states of consciousness, and emotional responses, serve as compelling evidence for the existence of ancient aliens in the eyes of many proponents of the Ancient Alien Theory. The consistency of these elements across various cultures and geographical locations suggests that these experiences may not be mere fabrications or the result of psychological phenomena, but rather genuine encounters with otherworldly beings who have been interacting with humanity for millennia.

The emotional impact of alien abduction and close encounter experiences can be complex and varied, encompassing feelings of fear, confusion, awe, and even a sense of connection or kinship with the extraterrestrial beings involved. These encounters often leave abductees with lingering questions about the purpose of their experiences, the intentions of the beings they encountered, and the possible implications for humanity's understanding of the universe.

Many abductees experience intrusive memories and flashbacks related to their encounters, which can be distressing and difficult to manage. These vivid recollections can lead to sleep disturbances, difficulty concentrating, and heightened sensitivity to stimuli that trigger reminders of the abduction experience. The persistence of these memories can contribute to the development of anxiety and PTSD symptoms, making it challenging for abductees to move past their encounters and resume a sense of normalcy in their lives.

For some individuals, the psychological effects of alien abduction and close encounter experiences can lead to profound changes in their sense of identity and personal growth. Abductees may begin to question their beliefs about the nature of reality, the existence of extraterrestrial life, and their place in the cosmos. These experiences can also inspire a newfound sense of curiosity, wonder, and open-mindedness, as abductees grapple with the implications of their encounters and explore the possibility of a broader, more interconnected universe.

Abductees often face stigmatization and social isolation as a result of their experiences, which can exacerbate the psychological challenges they are already dealing with. Friends, family members, and colleagues may struggle to understand or accept the abductee's account of their encounter, leading to feelings of rejection, isolation, and loneliness. This lack of support can make it difficult for abductees to process their experiences and find the help they need to cope with the psychological effects of their encounters.

Despite the significant psychological challenges associated with alien abduction and close encounter experiences, many abductees demonstrate remarkable resilience and adaptability. They develop various coping strategies to help them manage the emotional and cognitive effects of their encounters, such as engaging in creative expression, seeking support from others who have had similar experiences, or pursuing a deeper understanding of extraterrestrial life and the nature of the universe. By doing so, they are able to integrate their experiences into their lives and, in some cases, even find a sense of meaning and purpose in their encounters with otherworldly beings.

One of the most effective forms of support for abductees is the opportunity to connect with others who have had similar experiences. Peer support can help abductees feel less isolated and alone, providing them with a sense of community and belonging. Additionally, sharing stories and experiences with others who have undergone similar encounters can help abductees better understand and process their own experiences, fostering personal growth and healing.

In addition to peer support, many abductees benefit from professional support and counseling. Mental health professionals, such as psychologists and therapists, can help abductees address the emotional and psychological challenges they face as a result of their experiences. Through various therapeutic techniques, these professionals can assist abductees in developing coping strategies, building resilience, and improving their overall mental health and well-being.

Educating abductees about the phenomenon of alien abduction and close encounters can also play a crucial role in helping them cope with their experiences. Access to accurate and reliable information can empower abductees to better understand their encounters, the potential implications, and the broader context of ancient alien theory. This knowledge can help abductees make sense of their experiences and integrate them into their personal worldview, ultimately reducing feelings of confusion and fear.

For some abductees, spiritual and holistic support can provide valuable coping mechanisms and a sense of peace and purpose. Many abductees find solace and meaning in exploring their spiritual beliefs, whether that involves traditional religious practices or alternative spiritual paths. Practices such as meditation, energy healing, and mindfulness can also help abductees manage the emotional and psychological effects of their encounters, promoting inner peace and balance.

Advocacy and raising awareness about the needs and experiences of abductees can play a crucial role in promoting understanding and acceptance, as well as in the development of appropriate support services. By sharing their stories and educating others about the reality of alien abduction and close encounter experiences, abductees and their allies can help break down the stigma and skepticism that often surrounds this phenomenon. This can lead to increased societal recognition of the importance of providing comprehensive support and resources for those who have experienced encounters with otherworldly beings.

Overall, the provision of appropriate support and coping mechanisms for abductees is essential for their well-being and recovery. By combining peer support, professional counseling, education, spiritual practices, and advocacy, abductees can find the strength and resilience to navigate the complex emotional and psychological landscape that arises from their encounters with the unknown.

Hypnosis has proven to be an effective tool in helping abductees access and recover memories of their encounters with extraterrestrial beings. By inducing a relaxed, trance-like state, hypnosis allows individuals to bypass their conscious mind's defenses and delve into the deeper, subconscious layers of their memory. This process can reveal previously inaccessible details of abduction experiences, providing valuable insights into the nature of these encounters and the beings involved.

There are numerous accounts of abductees successfully recovering detailed memories of their alien encounters through the use of hypnotic regression. These recollections often contain specific and consistent information about the appearance of the extraterrestrial beings, their methods of communication, and the technology they employ. In some cases, the recovered memories have even led to the identification of physical evidence, such as unexplained scars or implants, that corroborate the abductees' accounts. These success stories lend credibility to the use of hypnosis as a means of investigating abduction experiences and further support the existence of ancient aliens.

One of the most compelling arguments in favor of the use of hypnosis in abduction investigations is the remarkable consistency of the narratives that emerge during hypnotic regression sessions. Abductees from diverse backgrounds and cultures often report strikingly similar experiences and details, suggesting that these memories may represent genuine encounters with extraterrestrial beings rather than the product of suggestion or false memory creation. This consistency lends credibility to the idea that hypnosis can be a valuable tool for uncovering the truth about alien abduction experiences and their connection to ancient aliens.

While it is true that hypnosis can be susceptible to suggestion and false memory creation, skilled hypnotists and abduction researchers are aware of these potential pitfalls and take measures to minimize their influence. By using open-ended questioning techniques and avoiding leading questions, hypnotists can reduce the likelihood of implanting false memories or influencing the abductees' accounts. Additionally, the process of corroborating the recovered memories with other evidence, such as physical marks or corroborating accounts from other witnesses, can help to ensure that the information obtained through hypnosis is reliable and accurate.

Despite the concerns surrounding the use of hypnosis in abduction investigations, it remains an important and valuable tool for exploring the phenomenon of alien abduction and its connection to ancient aliens. As researchers continue to refine their methods and improve their understanding of the psychological and neurological aspects of abduction experiences, hypnosis will likely remain a key component of this ongoing quest for knowledge and understanding. By carefully and responsibly employing hypnotic regression, abduction researchers can continue to uncover valuable insights into the nature of these encounters and their potential implications for our understanding of human history and our place in the universe.

Skeptics often attribute alien abduction experiences to sleep paralysis, a condition in which an individual is temporarily unable to move or speak while waking up or falling asleep. This phenomenon is often accompanied by vivid hallucinations, which could account for the vivid and bizarre imagery reported by abductees. However, proponents of the Ancient Alien Theory argue that sleep paralysis cannot explain the consistency of abduction experiences across cultures, the shared elements of these experiences, or the intricate details of alien technology and environments that abductees describe.

Another skeptical explanation for alien abduction and close encounter experiences is the power of suggestion. Critics suggest that exposure to stories, movies, or other cultural influences involving extraterrestrial contact could predispose individuals to interpret their experiences as encounters with aliens. While it is true that the power of suggestion can influence perception, Ancient Alien Theory proponents argue that this explanation fails to account for the striking similarities in abduction accounts from individuals with no prior knowledge of the phenomenon or exposure to similar stories.

Skeptics also maintain that the human mind has a tendency to create stories and explanations for unusual or inexplicable events, leading to the development of false memories or misinterpretations of real experiences. This argument, however, does not address the fact that many abduction accounts involve experiences that are highly consistent and detailed, often including descriptions of advanced technology and alien environments that are difficult to dismiss as mere fabrications of the human mind.

In some cases, abductees report physical evidence of their encounters, such as unexplained scars, marks, or implants. Skeptics may argue that these physical manifestations are the result of self-inflicted injuries or other mundane causes, but proponents of the Ancient Alien Theory contend that the presence of such evidence, in conjunction with the consistency of abduction experiences, lends credibility to the possibility of extraterrestrial contact.

While skeptical explanations for alien abduction and close encounter experiences offer alternative perspectives, they often fail to fully account for the complexity and consistency of these experiences. Proponents of the Ancient Alien Theory argue that the remarkable similarities in abduction accounts, the detailed descriptions of advanced technology and alien environments, and the presence of physical evidence in some cases, suggest that these experiences cannot be easily dismissed as mere hallucinations or products of the human mind's imagination. Instead, they argue that these encounters provide compelling evidence for the existence of ancient aliens and their ongoing interactions with humanity.

The connection between alien abduction experiences and Ancient Alien Theory is reinforced by the notion that these encounters could be evidence of ongoing contact between humans and extraterrestrial beings. By studying the details and patterns found in abduction accounts, proponents of Ancient Alien Theory argue that these experiences reveal a consistent and purposeful interaction between humans and otherworldly visitors. This interaction may be driven by various motives, such as the extraterrestrials' interest in our biological makeup, social and cultural development, or technological advancements.

Ancient Alien Theory suggests that extraterrestrial beings may have played a role in shaping human evolution and development. Alien abduction experiences, in which abductees often report undergoing medical examinations or procedures, could be seen as evidence of this influence. Proponents argue that these encounters may be part of a larger effort by extraterrestrials to monitor and guide our progress, ensuring that humanity follows a specific path or trajectory that aligns with their interests or objectives.

Ancient Alien Theory also proposes that extraterrestrials may have provided humans with advanced knowledge and technology throughout history. Alien abduction experiences, in which abductees sometimes report being exposed to advanced technologies or receiving information from their captors, could be considered evidence of this transfer of knowledge. This perspective suggests that abduction experiences serve as a means for extraterrestrials to share their wisdom and insights with humanity, potentially accelerating our technological and cultural development.

The prevalence of alien abduction accounts across cultures and throughout history lends further support to the connection between these experiences and Ancient Alien Theory. From ancient texts that describe encounters with otherworldly beings to modern-day reports of abductions, these accounts exhibit remarkable similarities despite the vast differences in time and cultural context. This cross-cultural consistency suggests that the abduction phenomenon may be a genuine manifestation of extraterrestrial contact, rather than a purely psychological or cultural construct.

Alien abduction experiences, when viewed through the lens of Ancient Alien Theory, can provide valuable insights into the nature and extent of extraterrestrial involvement in human history. By examining the patterns, commonalities, and potential motives underlying these experiences, we can gain a deeper understanding of the role that otherworldly visitors may have played in shaping our past and the ways in which they continue to influence our present. As a result, the phenomenon of alien abduction serves as an important and intriguing component of the broader Ancient Alien Theory, inviting further investigation and inquiry into the enduring question of our relationship with extraterrestrial life.

The biblical account of Enoch's ascent to the heavens is a prime example of an ancient narrative that appears to describe an alien abduction experience. In the Book of Enoch, the patriarch is taken by celestial beings known as the "Watchers" and shown various parts of the universe, as well as advanced knowledge that was beyond the understanding of humans at that time. The details of Enoch's journey and the knowledge he gained suggest the possibility of extraterrestrial contact, and proponents of ancient alien theory argue that this account serves as evidence of such encounters in our distant past.

Native American cultures have a rich tradition of stories involving otherworldly beings, often referred to as the Star People or Sky People. These beings are described as coming from the stars and interacting with humans in various ways, such as sharing knowledge, offering guidance, and even interbreeding with humans. The similarities between these stories and modern alien abduction accounts suggest a possible connection between ancient Native American experiences and extraterrestrial beings. Proponents of ancient alien theory argue that these stories further support the idea that our ancestors were visited and influenced by advanced extraterrestrial civilizations.

Ancient Hindu texts, such as the Vedas and the Mahabharata, contain numerous accounts of gods and celestial beings visiting Earth, interacting with humans, and demonstrating advanced knowledge and technology. For example, the Vimanas, described as flying chariots or aircraft, are said to have been used by gods to travel through the skies and even between different worlds. These accounts, along with descriptions of advanced weapons, aerial battles, and other remarkable feats, have led proponents of ancient alien theory to suggest that these texts provide evidence of extraterrestrial contact and intervention in human affairs.

The ancient Sumerian civilization also offers intriguing accounts of encounters with otherworldly beings, known as the Anunnaki. These beings were said to have come from the heavens to Earth and played a significant role in shaping human civilization, teaching advanced knowledge and even creating humans through genetic manipulation. The accounts of the Anunnaki bear a striking resemblance to modern alien abduction experiences and support the notion that human history has been influenced by extraterrestrial beings.

The historical accounts of alien abduction and close encounters found in various ancient texts, myths, and legends suggest a long-standing pattern of interaction between humans and otherworldly beings. While these accounts do not provide definitive proof of extraterrestrial contact, they offer compelling evidence that supports the ancient alien theory and its assertion that our ancestors were visited and influenced by advanced extraterrestrial civilizations. By examining these historical accounts, we can gain a deeper understanding of the possible connections between ancient human experiences and the phenomenon of alien abduction, and consider the implications of these encounters for our understanding of human history and our place in the universe.

The increasing number of alien abduction and close encounter experiences reported in recent decades has contributed significantly to the public discourse surrounding extraterrestrial life. These firsthand accounts, often shared through various media channels, have captured the attention of people from all walks of life, prompting widespread curiosity and speculation about the existence of advanced civilizations beyond our own. As a result, the phenomenon of alien abduction has helped to shift public perceptions of extraterrestrial life from a fringe topic to a more mainstream area of interest and debate.

The impact of abduction experiences on perceptions of extraterrestrial life extends beyond mere curiosity and speculation. These encounters have served as a rich source of inspiration for writers, filmmakers, and artists, who have used them to create compelling narratives and imaginative depictions of otherworldly beings and their advanced technologies. These creative works, in turn, have helped to further popularize the ancient alien theory and increase public awareness of the possibility of contact between humans and extraterrestrial civilizations.

Additionally, the growing interest in alien abduction and close encounter experiences has spurred scientific research into the search for extraterrestrial life, with organizations like the Search for Extraterrestrial Intelligence (SETI) dedicating resources to the detection of potential signals from other civilizations. This increased focus on the search for extraterrestrial life has led to significant advancements in our understanding of the cosmos and our ability to detect potential signs of life beyond Earth.

The phenomenon of alien abduction and its connections to ancient alien theory have also played a role in challenging traditional paradigms and encouraging open-mindedness about the nature of extraterrestrial life and the possibility of contact. By providing alternative explanations for historical events, archaeological discoveries, and cultural myths, proponents of ancient alien theory have encouraged people to question established beliefs and consider the possibility that our understanding of human history and the universe may be incomplete or inaccurate.

As the impact of abduction experiences on perceptions of extraterrestrial life continues to grow, so too does the potential for these experiences to shape the future of human-extraterrestrial relations. If contact between humans and otherworldly beings is indeed occurring, the accounts of abductees offer valuable insights into the nature of these interactions and the motivations of the beings involved. By examining these experiences and their potential implications, we may be better equipped to navigate future encounters with extraterrestrial civilizations and forge meaningful connections with our cosmic neighbors.

The significance of alien abduction and close encounter experiences extends beyond individual accounts and can provide us with valuable insights into the development of human civilization. Proponents of ancient alien theory argue that the knowledge and technology imparted by extraterrestrial visitors have played a crucial role in shaping the course of human history. By examining abduction experiences and their potential connections to ancient alien theory, we can explore the possibility that our ancestors benefited from advanced knowledge and guidance, which enabled them to build impressive structures, develop sophisticated technologies, and establish complex societies.

The phenomenon of alien abduction and its ties to ancient alien theory challenge the conventional narrative of human history, encouraging us to question the underlying assumptions about our origins and the development of our species. By considering the possibility of extraterrestrial contact and its influence on human history, we can open our minds to alternative explanations and interpretations that may reveal a more nuanced and complex understanding of our past.

Alien abduction experiences and their connections to ancient alien theory also offer a unique opportunity to explore the possibility of shared cultural experiences and the role of extraterrestrial contact in shaping our collective consciousness. The similarities in abduction accounts and ancient myths from various cultures around the world suggest a common thread that transcends geographical and cultural boundaries. This shared experience may indicate a universal human connection with extraterrestrial beings, further supporting the notion that our history is intertwined with that of otherworldly visitors.

The study of alien abduction experiences and their relation to ancient alien theory can also help expand our understanding of the cosmos and our place within it. By entertaining the possibility of extraterrestrial life and the potential for ongoing contact with otherworldly beings, we are compelled to consider our position in the grand scheme of the universe. This perspective can foster a sense of humility and curiosity, inspiring us to continue our search for answers about the origins of life, the potential for other advanced civilizations, and the mysteries of the cosmos that lie beyond our current understanding.

Lastly, examining the significance of alien abduction experiences and their connections to ancient alien theory may open the door to a greater future collaboration between humanity and extraterrestrial civilizations. By acknowledging the possibility of past interactions and learning from these experiences, we may be better equipped to engage in meaningful dialogue and cooperation with other advanced species in the future. This collaboration could lead to unprecedented advancements in technology, knowledge, and understanding, ultimately benefiting both humanity and our potential cosmic neighbors.

Chapter 10

Crop Circles, Animal Mutilation, and Other Unexplained Occurrences

A Phenomena Across the World

Over the years, crop circles have evolved from simple shapes and patterns to more elaborate and intricate designs. These formations often exhibit incredible symmetry and precision, incorporating complex geometric patterns and mathematical principles, such as the Golden Ratio and fractals. The increasing complexity of crop circles has led many to believe that their creation requires a level of knowledge and skill beyond that of ordinary human hoaxers.

In addition to their intricate designs, crop circles often exhibit a number of unexplained characteristics that suggest the involvement of advanced technology or non-human intervention. For example, the plants within some formations are not simply flattened but are bent at precise angles without breaking or damaging the stalks. Furthermore, some crop circles display evidence of exposure to intense heat or radiation, such as elongated and swollen plant nodes.

These peculiar characteristics are difficult to replicate through human-made methods or to attribute to natural phenomena, such as wind patterns or animal activity. As a result, proponents of the Ancient Alien Theory argue that these unexplained features provide evidence of extraterrestrial involvement in the creation of crop circles.

Many crop circles appear to contain encoded messages or symbols, which some researchers believe may be attempts at communication from extraterrestrial beings. Some of these formations incorporate binary code, mathematical equations, or symbolic representations of astronomical phenomena. For example, a crop circle discovered in Chilbolton, England, in 2001 appeared to contain a response to the Arecibo message, a radio signal transmitted into space by humans in 1974. The Chilbolton formation displayed a binary code that seemed to provide information about the purported extraterrestrial senders, including their DNA structure and the layout of their solar system.

These complex messages and symbols found in crop circles further support the idea that their creation may be the result of advanced intelligence and technology, potentially of extraterrestrial origin.

Crop circles often appear in close proximity to other unexplained phenomena, such as UFO sightings, mysterious lights in the sky, or areas with a history of paranormal activity. This correlation suggests a possible connection between these occurrences and the creation of crop circles. Proponents of the Ancient Alien Theory argue that these links provide additional evidence that extraterrestrial beings may be responsible for the formation of crop circles and that their appearance may be related to other instances of unexplained activity.

The mysterious patterns and designs found in crop circles, along with their unexplained characteristics, complex messages, and correlations with other phenomena, provide a compelling case for the involvement of ancient aliens in their creation. The precision, complexity, and seemingly encoded information within these formations point towards a level of intelligence and technology that is beyond our current understanding, further fueling the belief in the existence of extraterrestrial life and its interaction with our planet.

In some crop circle cases, the plants exhibit physiological changes that cannot be easily explained by conventional methods. One such change is the elongation and bending of plant nodes, which allow the plants to bend without breaking. This feature is not typically observed in instances where crop circles are created by human hands or simple mechanical means. Furthermore, these plants have been found to continue growing despite their bent and altered state, suggesting that whatever process is responsible for their formation does not cause lasting damage to the plant's cellular structure.

Some crop circle formations have been associated with elevated radiation levels and other electromagnetic effects. This has been observed in the form of increased background radiation in and around the formation, as well as unusual magnetic anomalies in the soil. These findings have led researchers to speculate that the forces responsible for creating these crop circles may be harnessing electromagnetic energy or other advanced technologies that are currently beyond our comprehension.

In certain crop circle cases, researchers have discovered the presence of unusual substances and particles within the affected areas. One notable example is the finding of microscopic iron spheres within the soil of some formations. These spheres are thought to be the result of intense heat or energy exposure, which could be indicative of an extraterrestrial technology at work. Additionally, changes in the crystalline structure of the soil have been observed in some instances, further supporting the notion that advanced energy sources may be involved in the creation of these enigmatic patterns.

Another compelling aspect of crop circles that lends credence to the involvement of ancient aliens is the geometric precision and mathematical patterns often found in these formations. Many crop circles display complex patterns that incorporate principles of sacred geometry, fractals, and other advanced mathematical concepts. It has been argued that the level of precision and sophistication required to create such patterns would be extremely difficult, if not impossible, for humans to achieve using conventional methods. This has led some researchers to propose that an advanced, non-human intelligence is responsible for the creation of these intricate designs.

There have been numerous instances where crop circles have appeared in conjunction with UFO sightings or other anomalous phenomena. In some cases, witnesses have reported seeing strange lights or objects in the sky immediately before or after the discovery of a crop circle. This correlation has led proponents of the Ancient Alien Theory to suggest that these events are not mere coincidences, but rather indicative of a connection between crop circles and extraterrestrial activity.

The scientific examination of crop circle formations has revealed a number of intriguing anomalies that defy conventional explanations. The physiological changes in plants, elevated radiation levels, presence of unusual substances and particles, geometric precision, and correlation with UFO sightings all point towards the possibility of extraterrestrial involvement. While further research is necessary to establish definitive proof, these findings provide compelling evidence in favor of the Ancient Alien Theory and its assertion that advanced, non-human intelligence may be responsible for these enigmatic patterns.

Animal mutilation cases have a number of perplexing characteristics that make them difficult to attribute to natural predators, disease, or human activity. In addition to the surgical precision and lack of blood, other common features include the absence of tracks or footprints around the mutilated animals and the presence of unusual puncture wounds or marks on the carcasses. Some reports even describe animals being discovered in locations that are difficult to access, suggesting that they may have been placed there by some external force.

The reasons behind the potential extraterrestrial involvement in animal mutilations remain a subject of speculation. Some researchers propose that the mutilations could be part of an ongoing effort to study terrestrial life forms, particularly in relation to genetics and reproduction. This could explain the frequent targeting of reproductive organs and the surgical precision with which they are removed. Other theories suggest that the mutilations could be a means of harvesting biological materials for unknown purposes or even part of a larger plan to manipulate Earth's ecosystem.

The advanced nature of the techniques used in animal mutilations points to the possibility of extraterrestrial technologies at work. For example, the surgical precision of the cuts, combined with the absence of blood, may indicate the use of advanced laser or electromagnetic tools that cauterize wounds as they are made. Similarly, the absence of tracks or footprints around the mutilated animals could suggest the use of levitation or transportation technologies that leave no trace of their presence.

The similarities between animal mutilation cases and reports of human abduction by extraterrestrial beings provide further support for the Ancient Alien Theory's involvement in these phenomena. Both types of incidents often involve the removal of tissue or organs with surgical precision, the presence of unusual marks or puncture wounds, and the experience of missing time or memory loss. These parallels suggest a shared origin for the phenomena and raise the possibility that the same extraterrestrial entities may be responsible for both animal mutilations and human abductions.

The ongoing and unexplained nature of animal mutilation cases serves as a compelling piece of evidence for proponents of the Ancient Alien Theory. The technological expertise and advanced knowledge displayed in these incidents point to the involvement of a highly advanced civilization, potentially with extraterrestrial origins. By considering the possibility of ancient aliens' involvement in animal mutilations, we can gain new insights into their motivations, methods, and the extent of their interactions with life on Earth.

In response to the increasing number of cattle mutilation reports in the 1970s, the United States government launched Operation Animal Mutilation, a federally funded investigation led by New Mexico State Police Officer Gabe Valdez. The aim of the investigation was to uncover the truth behind these mysterious incidents and determine if they were the result of natural predation, human activity, or something more unexplained.

During the same time, the U.S. Air Force also had an interest in the phenomenon, as they were concerned about potential links to UFO sightings and other unexplained aerial phenomena. This interest led to the creation of Project Grudge, a classified investigation into cattle mutilations and their potential connection to extraterrestrial activity.

Despite extensive investigations by law enforcement agencies and government-sponsored projects, the cattle mutilation phenomenon remains largely unsolved. While some cases have been attributed to natural predators or human actions, many others exhibit anomalies that cannot be easily explained by these factors. The failure to identify a definitive cause for these mutilations has fueled suspicions of a cover-up or suppression of information related to extraterrestrial activity.

Proponents of the Ancient Alien Theory argue that the government's interest in cattle mutilations and their inability to provide conclusive explanations may be indicative of a larger, hidden truth. They suggest that the unexplained nature of these cases points to the possibility of extraterrestrial involvement and that the government may be withholding information to prevent public panic or protect classified information related to advanced technology or extraterrestrial contact.

The involvement of government agencies in investigating cattle mutilations has been a source of intrigue for those who believe in the Ancient Alien Theory. Some argue that the government's interest in the phenomenon is an indication of its significance and the potential connection to extraterrestrial activity.

Furthermore, the lack of transparency in some of these investigations has led to claims of deliberate suppression or misinformation. For example, some researchers and whistleblowers have alleged that certain agencies, such as the FBI or the Department of Energy, have had direct knowledge of extraterrestrial involvement in cattle mutilations but have chosen to withhold this information from the public.

The idea that government agencies may be involved in a cover-up related to cattle mutilations and extraterrestrial activity has significant implications for our understanding of human history and our place in the universe. If extraterrestrial beings are indeed responsible for these mutilations, it suggests that there is a level of technology and knowledge far beyond our current understanding.

Furthermore, if the government is actively suppressing information regarding extraterrestrial involvement in cattle mutilations, it raises questions about the extent of human interaction with these beings and the potential existence of secret agreements or collaborations between the two parties.

The cattle mutilation phenomenon and the government's involvement in its investigation provide a compelling case for proponents of the Ancient Alien Theory. The lack of definitive explanations for these incidents and the potential suppression of information related to extraterrestrial activity continue to fuel speculation and debate about the true nature of these enigmatic occurrences and their implications for our understanding of the universe and our place within it.

In addition to mysterious lights and UFO sightings, various other aerial phenomena have been reported that defy conventional explanation. Witnesses have observed inexplicable formations of lights, glowing orbs, or strange objects that appear to move intelligently and display advanced capabilities. These occurrences, while not as well-known as classic UFO sightings, still contribute to the body of evidence that proponents of the Ancient Alien Theory believe points to extraterrestrial activity.

Around the world, people have reported experiencing unexplained sounds and vibrations that cannot be attributed to known natural or human-made sources. These strange sounds, often referred to as "skyquakes" or "mystery booms," have been described as loud, rumbling noises that seem to originate from the sky or the ground. Ancient Alien Theory proponents suggest that these unexplained phenomena could be the result of advanced extraterrestrial technology, possibly related to propulsion systems or communication efforts.

Throughout history, there have been numerous cases of mysterious disappearances and unexplained events that have left investigators baffled. Instances such as the Bermuda Triangle, where ships and planes have vanished without a trace, or the Dyatlov Pass incident, where a group of hikers in Russia died under mysterious circumstances, have captured the imagination of researchers and the public alike. Proponents of the Ancient Alien Theory argue that these enigmatic events may be connected to extraterrestrial activity, with alien beings or advanced technology playing a role in these baffling occurrences.

Some individuals who have reported encounters with extraterrestrial beings or UFO sightings have also claimed to discover unexplained physical evidence on their bodies, such as small metallic objects embedded beneath the skin, referred to as "alien implants." These objects, when surgically removed and analyzed, have displayed unusual properties and materials that are not easily explained by conventional science. Ancient Alien Theory proponents argue that these implants could be a form of extraterrestrial technology used for tracking, communication, or other unknown purposes, further supporting the idea of ongoing alien involvement in human affairs.

These various unexplained occurrences, including mysterious sky phenomena, unexplained sounds, disappearances, and physical evidence, all share common characteristics that suggest a possible connection to extraterrestrial activity. By examining these events through the lens of the Ancient Alien Theory, researchers can explore the potential implications of these phenomena and their potential link to advanced alien technology or intelligence.

While definitive proof of extraterrestrial involvement in these occurrences remains elusive, the anomalies and mysteries surrounding these phenomena continue to captivate the imagination and provide intriguing avenues for further investigation. As we continue to explore the unknown and seek answers to these enigmatic events, the Ancient Alien Theory will remain a compelling framework for understanding the potential connections between humanity and other intelligent life forms in the cosmos.

The phenomenon of mysterious lights and UFO sightings can be traced back to ancient times, with numerous historical accounts describing objects and lights in the sky that defy conventional explanations. Ancient texts and oral traditions from diverse cultures, such as the Egyptian, Sumerian, and Chinese civilizations, contain stories of celestial phenomena and flying objects that display characteristics consistent with modern UFO sightings. These ancient accounts suggest that the phenomenon of mysterious lights and UFOs has been a persistent feature of human history, further supporting the possibility of extraterrestrial visitation.

There have been numerous instances of UFO sightings and encounters reported by military personnel and documented in official records. These encounters often involve highly trained pilots and other military personnel who are well-versed in the identification of conventional aircraft and natural phenomena. In many cases, these UFOs have been observed to perform maneuvers and reach speeds that are beyond the capabilities of known human-made aircraft. The credibility of these witnesses and the official nature of the documentation lend further weight to the possibility that these sightings may be evidence of extraterrestrial activity.

To better understand and categorize the various types of UFO sightings and encounters, researchers have developed a classification system known as the Close Encounter system. This system, developed by astronomer and UFO researcher J. Allen Hynek, divides UFO encounters into three main categories: Close Encounters of the First Kind, where a UFO is observed at a distance; Close Encounters of the Second Kind, where a UFO has a physical effect on the environment or the witnesses; and Close Encounters of the Third Kind, where occupants of the UFO are observed or interact with witnesses.

This classification system provides a structured framework for analyzing and understanding the wide range of UFO sightings and encounters, offering further insight into the nature and characteristics of these mysterious phenomena. By examining these encounters through the lens of the Ancient Alien Theory, researchers can explore the potential connections between these sightings and extraterrestrial visitation.

Proponents of the Ancient Alien Theory argue that the global and historical nature of UFO sightings and mysterious lights in the sky supports the possibility of extraterrestrial visitation. These sightings often share common characteristics, such as the advanced technology and capabilities displayed by the observed objects, which are beyond the scope of human invention. Furthermore, the persistence of these sightings across different cultures and time periods suggests that this phenomenon is not merely the product of human imagination or misidentification of natural phenomena.

In addition to the sightings themselves, UFO researchers have identified patterns and correlations that may indicate a connection to extraterrestrial activity. For example, UFO sightings have been found to occur more frequently in areas with high concentrations of nuclear facilities, military installations, and technological research centers. This pattern suggests that extraterrestrial beings may be monitoring or even influencing human technological development, possibly out of concern for the potential consequences of our advancements.

The study of mysterious lights and UFO sightings offers valuable insights into the potential existence of ancient aliens and their ongoing interactions with our planet. While the precise nature of these phenomena remains a subject of debate, the wealth of historical accounts, credible witness testimonies, and official documentation provides compelling evidence for the possibility of extraterrestrial visitation. By examining these sightings through the lens of the Ancient Alien Theory, we may be able to gain a greater understanding of our place in the cosmos and the potential connections between humanity and other intelligent life forms in the universe.

The phenomenon of time loss involves individuals experiencing unaccounted-for periods, during which they have no recollection of events or actions that transpired. These memory gaps can range from minutes to hours or even days. Often, the affected individuals exhibit disorientation and confusion, struggling to piece together the events that occurred during the missing time. Proponents of the Ancient Alien Theory argue that these incidents may be linked to encounters with extraterrestrial beings, who may possess advanced technologies capable of manipulating human perception and memory.

In some cases of time loss, individuals also report the appearance of physical symptoms or unexplained marks on their bodies. These markings can include bruises, scratches, or more intricate patterns such as geometric shapes. Sometimes, the affected individuals experience lingering physical sensations, like soreness or numbness, in areas where these marks appear. Supporters of the Ancient Alien Theory propose that these marks and symptoms may be the result of extraterrestrial experimentation or contact, providing tangible evidence of these otherworldly encounters.

Many individuals who experience time loss also report having strange dreams or visions during or following the incident. These dreams often involve vivid imagery of unfamiliar environments, advanced technology, or interactions with beings that appear non-human. Some people even recall undergoing medical procedures or participating in experiments during these dream-like states. Ancient Alien Theory proponents argue that these dreams and visions may represent fragmented memories of actual encounters with extraterrestrial beings, offering glimpses into the experiences that transpired during the missing time.

One possible explanation for time loss and related phenomena is the abduction hypothesis, which suggests that individuals experiencing missing time may have been taken aboard extraterrestrial craft for examination or experimentation. This theory proposes that the advanced technology and capabilities of these otherworldly beings allow them to erase or manipulate human memories, resulting in the perception of time loss. The physical symptoms, unexplained marks, and strange dreams reported by affected individuals may be residual effects of these encounters or the body's response to the advanced technology used by the extraterrestrials.

The phenomenon of time loss and its potential connection to extraterrestrial activity offers a compelling area of inquiry for Ancient Alien Theory supporters. By examining the experiences of those who report missing time, researchers may gain valuable insights into the nature of extraterrestrial interactions with humans and the advanced technologies employed by these beings. Furthermore, these investigations may shed light on the broader implications of extraterrestrial visitation, such as the impact on human psychology, our understanding of the universe, and our place within the cosmos.

The unexplained phenomena discussed in this chapter, such as crop circles, animal mutilations, mysterious lights, UFO sightings, and time loss experiences, consistently demonstrate a level of advanced technology and knowledge beyond human capabilities. These instances often involve intricate designs, surgical precision, and anomalous characteristics that defy conventional explanations. Proponents of the Ancient Alien Theory contend that these occurrences serve as evidence of extraterrestrial intervention in our world, as the complex nature of these events suggests a technological prowess that surpasses human understanding.

Several unexplained phenomena exhibit elements of communication or interaction with humans. For example, the intricate designs found in crop circles have been interpreted as encoded messages or symbols that may represent an attempt at communication by extraterrestrial beings. Similarly, some UFO sightings have been reported to occur in response to human-made signals or appear to demonstrate a level of awareness of human presence.

This aspect of communication and interaction suggests that extraterrestrial beings may be actively engaging with humanity, perhaps seeking to understand our species, share knowledge, or even guide our development. By examining these occurrences through the lens of the Ancient Alien Theory, we may gain insight into the intentions, methods, and motivations of these otherworldly beings.

The connection between unexplained phenomena and the Ancient Alien Theory also raises the possibility that extraterrestrial beings have played a more direct role in shaping human civilization. From the construction of ancient megalithic structures to the development of advanced technology, these occurrences suggest that there may have been collaboration between humans and extraterrestrial beings throughout history.

This collaboration could extend to various aspects of human culture, including art, religion, and mythology. Many ancient myths and legends describe encounters with otherworldly beings or feature stories that can be interpreted as depicting extraterrestrial technology or intervention. These narratives, along with the presence of unexplained phenomena, lend credence to the idea that extraterrestrial beings have influenced the course of human history and continue to play a role in our world today.

Ultimately, the connection between unexplained phenomena and the Ancient Alien Theory can help expand our understanding of the universe and our place within it. By considering the possibility that extraterrestrial beings have interacted with our planet and its inhabitants, we open the door to new perspectives on the nature of intelligent life, the potential for interstellar communication and travel, and the shared experiences of sentient beings across the cosmos.

Exploring the connection between unexplained phenomena and the Ancient Alien Theory offers a unique perspective on the potential existence of ancient aliens and their ongoing interactions with our planet. The anomalies and mysteries surrounding these phenomena continue to captivate researchers, enthusiasts, and the general public alike. By examining these unexplained events through the lens of the Ancient Alien Theory, we may be able to gain a greater understanding of our place in the cosmos and the potential connections between humanity and other intelligent life forms in the universe.

In conclusion, the study of crop circles, animal mutilations, and other unexplained phenomena presents a compelling opportunity to explore the potential existence of ancient aliens and their ongoing interactions with our planet. By embracing the Ancient Alien Theory perspective, we can delve deeper into the anomalies and mysteries that surround these enigmatic occurrences.

As researchers and enthusiasts continue to investigate these phenomena, new discoveries and insights may come to light, further challenging our understanding of the world around us. The possibility of extraterrestrial involvement in these events, while not yet definitively proven, invites us to entertain a broader and more expansive view of the universe and our place within it.

By examining these unexplained phenomena through the lens of the Ancient Alien Theory, we can begin to explore the potential connections between humanity and other intelligent life forms in the universe. This perspective not only offers a fresh approach to understanding our own history but also encourages us to consider the profound implications of a universe teeming with intelligent life.

In the pursuit of knowledge and understanding, the Ancient Alien Theory fosters curiosity, open-mindedness, and the willingness to question established beliefs. By considering the possibility of ancient aliens and their potential influence on our planet, we can expand our horizons and deepen our appreciation for the mysteries of the universe. This perspective not only enriches our understanding of our own history and the natural world but also inspires us to seek out new discoveries and experiences, ultimately contributing to a more profound sense of wonder and awe in the face of the unknown.

In the end, the study of crop circles, animal mutilations, and other unexplained phenomena through the lens of the Ancient Alien Theory serves as a powerful reminder that our understanding of the universe is ever-evolving. As we continue to explore the mysteries of our world and our place within the cosmos, the potential connections between humanity and extraterrestrial life will undoubtedly remain a captivating and thought-provoking area of inquiry.

Chapter 11

Planetary Engineering and Terraforming

The Ancient Aliens and Our Modern Day

Planetary engineering and terraforming involve the manipulation of a planet's environment to make it more hospitable for life. While these concepts have long been the subject of science fiction, recent advances in technology have brought them closer to reality. In the context of Ancient Alien Theory, some researchers propose that extraterrestrial beings may have employed these techniques in Earth's past to create suitable conditions for life, thus playing a significant role in the development of our planet and its inhabitants.

The Cambrian Explosion, a period of rapid diversification of multicellular life forms around 541 million years ago, remains one of the most significant and mysterious events in Earth's history. Ancient Alien Theory proponents argue that the sudden appearance of a diverse array of species during this relatively short period may be indicative of extraterrestrial intervention. They suggest that ancient aliens could have played a role in seeding Earth with new life forms or manipulating the genetic makeup of existing organisms to accelerate their evolution and diversification.

The end of the last Ice Age, which occurred around 11,700 years ago, marked a dramatic shift in Earth's climate and a rapid rise in global temperatures. This change led to the retreat of ice sheets and the emergence of new ecosystems, paving the way for the rise of human civilization. Ancient Alien Theory proponents propose that this sudden climatic shift may have been influenced by the actions of advanced extraterrestrial beings. These ancient aliens could have employed advanced technology to manipulate Earth's climate, either by redirecting solar radiation or by influencing other factors that contribute to global temperature fluctuations.

The Great Oxygenation Event, which occurred approximately 2.4 billion years ago, marked the beginning of a significant increase in atmospheric oxygen levels. This process, which allowed for the development of complex life forms, is considered one of the most crucial events in Earth's history. Ancient Alien Theory researchers suggest that the oxygenation of Earth's atmosphere may have been the result of ancient alien engineering. By introducing photosynthetic organisms or employing other advanced methods, these extraterrestrial beings could have altered Earth's atmosphere, making it more hospitable for the evolution of complex life.

In addition to these major events, proponents of Ancient Alien Theory also point to various geological anomalies as potential evidence of ancient alien involvement in Earth's history. These include the presence of unusual mineral deposits, the formation of striking geological features such as the Richat Structure in Mauritania, and the existence of massive underground cave systems that remain largely unexplained. By examining these anomalies and their potential connections to extraterrestrial intervention, Ancient Alien Theory researchers hope to uncover further evidence of planetary engineering by advanced beings in our planet's distant past.

Ancient Alien Theory proposes that the remarkable and rapid transformations in Earth's history were significantly influenced by ancient aliens possessing advanced technology and knowledge beyond human understanding. These extraterrestrial beings, with their superior capabilities, may have engineered Earth's environment in ways that fostered the development of life and facilitated its flourishing.

By intervening in Earth's natural processes, ancient aliens may have guided the evolutionary path of life on our planet. They could have influenced the development of various species, directing their adaptations and diversification to create a rich and complex biosphere. This intervention might have been responsible for the rapid emergence of diverse life forms during events such as the Cambrian Explosion, as well as the unexpected survival and evolution of certain species following mass extinction events.

Some proponents of Ancient Alien Theory suggest that extraterrestrial beings may have seeded Earth with life, either by introducing simple organisms or by accelerating the process of abiogenesis. The panspermia hypothesis, which proposes that life on Earth originated from microorganisms or chemical precursors present in outer space, lends support to this idea. Ancient aliens may have utilized their advanced technology to transport the building blocks of life across the cosmos, ensuring the development of life on Earth and perhaps other planets.

Ancient Alien Theory researchers also propose that extraterrestrial beings may have been responsible for significant alterations in Earth's atmosphere and climate. By manipulating the composition of Earth's atmosphere, ancient aliens could have created conditions conducive to life, such as the oxygen-rich environment necessary for the development of complex organisms. Additionally, these beings may have controlled Earth's climate to maintain hospitable temperatures and weather patterns, thus ensuring the survival and prosperity of life on our planet.

The establishment of diverse and balanced ecosystems on Earth could be another result of ancient alien intervention. By introducing various species and guiding their development, ancient aliens may have created intricate ecological networks that have supported life throughout Earth's history. The striking similarities between Earth's ecosystems and those of other planets, as well as the presence of extremophiles that can survive in harsh conditions, might indicate extraterrestrial involvement in the development of life across the cosmos.

The idea that ancient aliens played a significant role in shaping Earth's environment and fostering the development of life remains a captivating aspect of Ancient Alien Theory. While concrete evidence of extraterrestrial intervention in planetary engineering has yet to be discovered, the rapid and remarkable changes in Earth's history, combined with the intriguing possibilities suggested by modern scientific research, continue to inspire curiosity and debate on this fascinating topic. As we continue to unravel the mysteries of our planet's past, the notion of ancient alien involvement in Earth's development will remain an intriguing area of inquiry for researchers and enthusiasts alike.

Water is indeed an essential component for life as we know it, providing the necessary environment for critical biological processes and acting as a solvent for a wide range of biochemical reactions. The presence of water on Earth, particularly in liquid form, has shaped the development of our planet's biosphere and has allowed for the evolution of a diverse array of life forms.

Ancient Alien theorists postulate that the distribution of water on Earth might not be solely the result of natural processes but could have been influenced by extraterrestrial intervention. They argue that ancient aliens, possessing advanced technology and knowledge, could have manipulated the planet's water resources to create the ideal conditions for life to emerge and thrive.

One intriguing piece of evidence cited by proponents of this theory is the presence of vast underground reservoirs, such as those discovered beneath the Sahara Desert. These reservoirs, which contain significant amounts of water, seem unlikely to have formed through natural processes alone, given the arid climate of the region. Ancient Alien theorists propose that these reservoirs are remnants of ancient alien engineering projects aimed at storing and distributing water across the planet, thus ensuring the availability of this vital resource for the development of life.

Beyond the underground reservoirs beneath the Sahara Desert, Ancient Alien theorists also point to other examples of what they believe to be ancient alien engineering projects related to water distribution. These include the construction of vast networks of aqueducts and canals in ancient civilizations, such as those in Mesopotamia and the Indus Valley, which may have been built with the guidance or assistance of extraterrestrial beings. The advanced knowledge and technology required to create these systems suggest that ancient humans may have had access to information and resources beyond their own capabilities, pointing to the possibility of alien intervention.

The potential influence of ancient aliens on Earth's water distribution could have had a profound impact on the development of life on our planet. By ensuring the availability of water in key regions, these extraterrestrial beings may have played a crucial role in the emergence of diverse ecosystems and the evolution of various species. This idea lends further support to the Ancient Alien Theory, suggesting that our planet's history and the development of life on Earth might have been shaped, in part, by the actions of intelligent extraterrestrial beings.

Ancient Alien Theory proponents suggest that advanced extraterrestrial beings possessed the knowledge and technology necessary to manipulate Earth's climate. This could have involved the use of advanced atmospheric and meteorological technologies to control temperature, humidity, and precipitation. By carefully regulating these factors, ancient aliens may have been able to promote the growth of specific ecosystems and encourage the evolution of particular species, ultimately shaping the development of life on Earth.

Numerous ancient myths and legends from various cultures around the world describe gods and other supernatural beings with the power to control the weather. These include the Greek god Zeus, who wielded lightning bolts; the Norse god Thor, who brought thunderstorms; and the Aztec god Tlaloc, who controlled rain and drought. Ancient Alien Theory proponents argue that these accounts could be interpreted as evidence of advanced beings with the ability to manipulate Earth's climate.

Some researchers believe that certain ancient structures may have been designed or influenced by extraterrestrial beings with the purpose of controlling Earth's climate. For example, the massive pyramids of Egypt and Mesoamerica, the enigmatic stone spheres of Costa Rica, and the intricate stone circles of Stonehenge have all been proposed as potential climate control devices built or inspired by ancient aliens. These structures, often featuring precise alignments with celestial bodies and demonstrating an advanced understanding of astronomy, may have functioned as tools for managing Earth's weather patterns and promoting specific environmental conditions.

The ability to control Earth's climate would have had a profound impact on the development of human civilization. By influencing weather patterns and ecosystems, ancient aliens could have guided the growth of agriculture, the rise and fall of empires, and the migrations of populations across the globe. By shaping the environment to suit their needs, these advanced beings may have played a critical role in the evolution of human societies and the course of our planet's history.

As our understanding of climate science grows, we have begun to develop our own methods of controlling Earth's weather, such as cloud seeding and geoengineering. Ancient Alien Theory proponents argue that our current efforts in climate control may have been influenced, either directly or indirectly, by the advanced techniques employed by ancient extraterrestrial beings. Studying the potential evidence of ancient alien climate control could provide valuable insights into our own efforts to manage Earth's climate and preserve our planet's delicate ecosystems.

The idea that ancient aliens may have engaged in climate control to shape Earth's environment is a compelling aspect of Ancient Alien Theory. While conclusive evidence remains elusive, the accounts of weather-controlling gods, mysterious ancient structures, and the potential influence on human civilization provide a fascinating foundation for further exploration. As we continue to unravel the mysteries of Earth's climate and develop our own methods of weather manipulation, the possibility of ancient alien involvement in shaping our planet's weather patterns will likely remain an intriguing and thought-provoking area of research.

The origins of life on Earth remain a subject of debate among scientists. While various theories have been proposed, ranging from abiogenesis to panspermia, none can fully explain the rapid development and diversification of life on our planet. Ancient Alien theorists suggest that the emergence of life on Earth may have been influenced or accelerated by the terraforming efforts of ancient aliens. These extraterrestrial beings, possessing advanced knowledge and technology, could have intentionally introduced the necessary building blocks for life, such as amino acids and RNA molecules, to jumpstart the process of evolution.

One crucial aspect of terraforming involves the manipulation of a planet's atmosphere to create a suitable environment for life. Ancient Alien proponents argue that the gradual shift in Earth's atmosphere, from a reducing environment rich in methane and ammonia to an oxygen-rich environment, was the result of deliberate intervention by ancient aliens. By introducing photosynthetic organisms, such as cyanobacteria, these extraterrestrial beings may have initiated the Great Oxygenation Event, paving the way for the development of more complex, oxygen-dependent life forms.

In addition to the manipulation of Earth's atmosphere, Ancient Alien theorists propose that these beings may have influenced other environmental factors, such as temperature, ocean currents, and nutrient distribution, to create optimal conditions for the evolution of life. For example, they point to the rapid warming during the Late Heavy Bombardment period, which allowed for the formation of liquid water oceans and the subsequent emergence of life. These changes, which occurred over relatively short geological time scales, could be indicative of ancient alien intervention and terraforming efforts.

Some proponents of Ancient Alien Theory argue that these extraterrestrial beings not only influenced Earth's environment but also directly intervened in the development of life through genetic engineering. By manipulating the genetic material of early life forms, ancient aliens may have guided the course of evolution, leading to the emergence of diverse and complex organisms, including the ancestors of modern humans. This idea is supported by the presence of so-called "junk DNA" in our genomes, which some researchers speculate could be the remnants of ancient alien genetic manipulation.

The potential involvement of ancient aliens in the terraforming of Earth and the development of life has profound implications for our understanding of human history and our place in the cosmos. By shaping the environment and guiding the evolution of life on our planet, these extraterrestrial beings may have played a crucial role in the rise of human civilization. This idea challenges conventional theories about our origins and encourages us to consider the possibility that our development has been influenced by intelligent life from beyond our planet.

The Permian-Triassic extinction event, which occurred approximately 252 million years ago, is the largest mass extinction in Earth's history. It is estimated that around 90% of marine species and 70% of terrestrial species went extinct during this period. Proponents of Ancient Alien Theory suggest that this extinction event may have been influenced by extraterrestrial beings, who could have employed advanced technology to manipulate Earth's environment and trigger such a catastrophic event. This could have been done to reshape Earth's ecosystems, allowing for the emergence of new life forms better suited for the planet's new conditions or to further their own terraforming efforts.

The Cretaceous-Paleogene extinction event occurred about 66 million years ago, leading to the extinction of approximately 75% of all species on Earth, including the non-avian dinosaurs. The prevailing scientific explanation for this event is the impact of a massive asteroid or comet, which triggered a series of catastrophic environmental changes. However, some Ancient Alien Theory researchers argue that this impact could have been orchestrated by ancient aliens, either as a means of controlling the development of life on Earth or as a side effect of their planetary engineering activities. By eliminating the dominant species of the time, these extraterrestrial beings could have opened up ecological niches for the rise of new life forms, including mammals and eventually humans.

The motivations behind potential ancient alien involvement in mass extinctions are a subject of much speculation. Some proponents of Ancient Alien Theory believe that these beings may have intentionally caused these catastrophic events to manipulate the course of life on Earth, shaping ecosystems and promoting the evolution of specific species to suit their own purposes or interests. Others propose that the mass extinctions were an unintended consequence of their planetary engineering efforts, with the ancient aliens possibly unaware of the full impact their actions would have on Earth's biosphere.

While there is no definitive evidence directly linking ancient aliens to Earth's mass extinction events, proponents of Ancient Alien Theory point to various factors that could suggest extraterrestrial involvement. These include the sudden and dramatic nature of these extinctions, the presence of unexplained geological and climatic anomalies during these periods, and the rapid emergence of new life forms in the aftermath of these events. Additionally, ancient myths and legends from various cultures around the world describe cataclysmic events and divine interventions that could be interpreted as evidence of ancient alien influence on Earth's history.

The idea that ancient aliens may have played a role in Earth's mass extinctions has significant implications for our understanding of our planet's history and the development of life. If true, this would suggest that the course of evolution on Earth has been influenced by external forces beyond our current comprehension, with the potential involvement of advanced extraterrestrial beings in shaping the ecosystems and life forms we see today. This perspective challenges conventional scientific explanations for Earth's history and encourages further research into the possible connections between ancient aliens and the major events that have shaped our world.

Modern terraforming efforts are often seen as a testament to humanity's ingenuity and technological prowess. However, proponents of Ancient Alien Theory suggest that these endeavors might have their roots in knowledge passed down from ancient extraterrestrial beings. They argue that the advanced techniques and principles used in today's terraforming projects could have originated from ancient alien civilizations that had already mastered the art of transforming inhospitable environments into thriving ecosystems.

Supporters of Ancient Alien Theory point to various instances in Earth's history as potential examples of ancient alien terraforming. Rapid and dramatic changes in our planet's environment, such as the oxygenation of the atmosphere and the sudden end of the Ice Age, may have been the result of extraterrestrial intervention. By studying these events, scientists may gain valuable insights into the methods used by ancient aliens to manipulate and shape Earth's environment, which could then be applied to modern terraforming projects.

As human civilization continues to advance and explore the cosmos, we have made significant strides in our understanding of the universe and our place within it. Ancient Alien Theory proponents argue that much of this progress can be attributed, at least in part, to the influence of extraterrestrial knowledge. They believe that ancient aliens may have provided our ancestors with advanced scientific knowledge, which was then passed down through generations and contributed to the development of modern science and technology. As a result, our current understanding of planetary engineering and terraforming could be deeply rooted in this ancient alien wisdom.

Looking to the future, the potential colonization of other planets such as Mars will likely require extensive terraforming efforts to create habitable environments. Ancient Alien Theory supporters contend that as we develop new techniques and technologies for this purpose, we may continue to uncover evidence of ancient alien influence in our scientific pursuits. These discoveries could further validate the idea that our modern understanding of terraforming is grounded in extraterrestrial knowledge and expertise.

As humanity pushes the boundaries of our understanding and capabilities in the realm of planetary engineering and terraforming, Ancient Alien Theory enthusiasts propose that we may one day encounter extraterrestrial life forms with similar goals. If such a meeting were to occur, it could pave the way for collaboration between our species and advanced alien civilizations, enabling us to learn from their vast knowledge and experience in planetary engineering. This exchange could lead to significant advancements in our ability to shape and transform the environments of other planets, ensuring the long-term survival and prosperity of our species.

The idea that modern terraforming efforts are influenced by ancient alien knowledge and techniques adds a fascinating dimension to the study of Ancient Alien Theory. As we continue to explore the potential for transforming other planets into habitable environments, we may uncover further evidence of extraterrestrial intervention in our planet's history, and perhaps even establish contact with advanced alien civilizations that share our interest in planetary engineering.

The Nazca Lines in Peru are a collection of large-scale geoglyphs that stretch across the Nazca Desert. These intricate designs, which include animal figures and geometric shapes, can only be fully appreciated from an aerial perspective. Some proponents of Ancient Alien Theory suggest that the Nazca Lines were created as a system for managing water resources, with the geoglyphs serving as markers for an ancient network of underground channels or wells. The scale and complexity of these designs, combined with their mysterious purpose, hint at the possibility of extraterrestrial guidance or involvement in their creation.

The ancient Sumerians were known for their advanced agricultural and irrigation techniques, which allowed them to cultivate fertile lands in the otherwise arid environment of Mesopotamia. These systems included canals, dams, and reservoirs that were engineered with remarkable precision and efficiency. Proponents of Ancient Alien Theory argue that the complexity and sophistication of these irrigation systems may be indicative of extraterrestrial guidance, as the Sumerians seemingly possessed knowledge and technology beyond what would be expected for their time.

Many megalithic structures found around the world, such as Stonehenge in England and the Great Pyramid of Giza in Egypt, have long puzzled researchers due to the precision and advanced engineering techniques used in their construction. Ancient Alien Theory proponents suggest that these structures may be evidence of extraterrestrial involvement in planetary engineering, with the advanced knowledge and technology required for their creation having been provided by ancient aliens. Additionally, some researchers have posited that these structures may have served a purpose in stabilizing Earth's environment or influencing its energy grid, further supporting the idea of ancient alien intervention in planetary engineering.

Evidence of ancient mining and resource extraction activities can be found across the globe, with some sites dating back thousands of years. These mining operations often targeted valuable minerals and metals, such as gold and copper, which would have been essential for the development of advanced technologies. Ancient Alien Theory proponents argue that the scale and sophistication of these ancient mining operations suggest that extraterrestrial beings may have been involved in their planning and execution, providing the necessary knowledge and technology to locate and extract these valuable resources.

Throughout history, numerous examples of advanced technologies have been discovered, many of which seem to be beyond the capabilities of the civilizations that created them. These include the Antikythera mechanism, an ancient Greek astronomical device with remarkable precision and complexity, and the Baghdad Battery, a possible ancient electrical device. Proponents of Ancient Alien Theory argue that these advanced technologies may be evidence of extraterrestrial involvement in planetary engineering, as they could have been used to manipulate Earth's environment or facilitate resource extraction and management efforts.

One of the most compelling aspects of the Mahabharata that supports the Ancient Alien Theory is the description of Vimanas, flying machines that were used by various characters in the epic. Vimanas were described as being powered by advanced propulsion systems, with the ability to travel vast distances and perform complex aerial maneuvers. The detailed descriptions of these flying machines, their construction, and their capabilities suggest that ancient civilizations may have had access to advanced technology that was beyond their own understanding. Ancient Alien proponents argue that this knowledge could have been provided by extraterrestrial beings, who may have also played a role in the construction and operation of the Vimanas.

The Mahabharata also contains accounts of advanced medical procedures and knowledge that seem to be beyond the capabilities of ancient civilizations. For example, the text describes complex surgeries, including the transplantation of limbs and organs, which were performed with remarkable precision and success. Additionally, the Mahabharata speaks of advanced genetic engineering techniques, such as the creation of hybrid creatures and the manipulation of human fertility. Ancient Alien theorists suggest that this advanced medical knowledge could have been provided by extraterrestrial beings, who may have shared their technology and expertise with ancient humans to advance the development of civilization.

The Mahabharata describes several ancient cities that were constructed with remarkable architectural and engineering prowess. Some of these cities, such as Dwarka and Hastinapur, were said to have been designed and built under the guidance of divine beings. The advanced technologies and materials used in the construction of these cities, as well as the complex urban planning and infrastructure systems that they employed, are indicative of knowledge that was beyond the capabilities of ancient humans.

Ancient Alien proponents argue that this expertise may have been provided by extraterrestrial beings, who played a role in the development of these advanced cities and their infrastructure.

The Mahabharata also contains extensive knowledge of astronomy and cosmology, including descriptions of celestial phenomena and the movements of celestial bodies. Some of the astronomical knowledge presented in the epic, such as the accurate prediction of solar and lunar eclipses, was remarkably advanced for its time. Ancient Alien theorists propose that this knowledge may have been provided by extraterrestrial beings, who shared their understanding of the cosmos with ancient humans to further the development of science and technology.

The advanced technologies, medical knowledge, architectural and engineering feats, and astronomical understanding described in the Mahabharata provide a compelling case for the Ancient Alien Theory. The epic's accounts of advanced weapons, flying machines, and divine intervention in human affairs suggest that ancient civilizations may have had contact with extraterrestrial beings, who shared their knowledge and expertise to shape the course of human history. By examining the Mahabharata and its wealth of information, researchers can gain a deeper understanding of the potential connections between ancient civilizations and extraterrestrial beings, shedding light on the origins of human civilization and the possibility of ancient alien involvement in our planet's development.

The Anunnaki, as described in the Sumerian creation myth, were part of a larger pantheon of gods who played critical roles in the development of human civilization. These gods were said to have not only shaped the Earth and its environment but also directly intervened in the lives of humans, teaching them various skills, arts, and sciences. Some Ancient Alien theorists argue that these gods were actually advanced extraterrestrial beings who provided humans with the knowledge and technology needed to establish complex societies and develop agriculture, ultimately leading to the rise of civilization.

In addition to their potential role in planetary engineering, some proponents of Ancient Alien Theory propose that the Anunnaki may have engaged in genetic engineering to create modern humans. According to this hypothesis, the Anunnaki combined their own DNA with that of hominids native to Earth, resulting in the creation of Homo sapiens. This idea is supported by the Sumerian myth of the creation of humans, which describes the gods forming humans from clay, a metaphor that could represent the process of genetic manipulation. If true, this would imply that the Anunnaki were not only responsible for shaping Earth's environment but also directly influenced the development of human civilization through genetic engineering.

The Sumerian creation myth credits the Anunnaki with teaching humans the art of agriculture, which played a crucial role in the development of civilization. Ancient Alien theorists suggest that this knowledge may have been imparted by extraterrestrial beings, who provided humans with the advanced techniques and tools needed to cultivate crops and domesticate animals. The development of agriculture enabled humans to settle in one place and form complex societies, which in turn led to the growth of cities and the establishment of powerful civilizations like ancient Mesopotamia.

In addition to their roles in genetic engineering and agriculture, the Anunnaki were also associated with various advanced technologies in Sumerian mythology. For example, they were said to possess flying machines known as "chariots of the gods," which allowed them to travel vast distances and communicate with humans. Some Ancient Alien theorists propose that these chariots were actually advanced spacecraft used by extraterrestrial beings to traverse the cosmos and visit Earth. If true, this would suggest that the Anunnaki were not only responsible for shaping Earth's environment but also provided humans with advanced technology that significantly influenced the course of human history.

The Sumerian civilization, which flourished in ancient Mesopotamia, left a lasting legacy that continues to influence modern society. From the development of written language to advancements in mathematics and astronomy, the Sumerians made numerous contributions that laid the groundwork for later civilizations. Ancient Alien theorists argue that the advanced knowledge and capabilities attributed to the Anunnaki in Sumerian texts indicate the possibility of extraterrestrial involvement in the development of this influential civilization. By exploring the connections between the Anunnaki, the Sumerian creation myth, and the various advanced technologies and skills attributed to these beings, researchers can gain a better understanding of the potential role of ancient aliens in shaping human history.

The Watchers, as described in the Book of Enoch, were said to have imparted advanced knowledge to humanity that would have been difficult for ancient civilizations to acquire on their own. This knowledge included metallurgy, which allowed humans to create sophisticated tools and weapons; agriculture, which played a crucial role in the development of human societies; and astronomy, which provided insight into celestial events and the movement of celestial bodies. Ancient Alien theorists argue that the complexity of these skills and the rapid development of human civilizations can be better explained by the intervention of extraterrestrial beings than by natural human progress alone.

In addition to providing advanced knowledge, some proponents of Ancient Alien Theory suggest that the Watchers may have been involved in genetic manipulation to accelerate human evolution. The Book of Enoch describes the Watchers interbreeding with human women, resulting in the birth of giants known as the Nephilim. This interbreeding could be interpreted as an attempt by extraterrestrial beings to enhance human genetics and further the development of our species. The existence of the Nephilim and their unusual characteristics could be viewed as evidence of genetic experimentation by ancient aliens.

The Book of Enoch also describes the Watchers' ability to manipulate Earth's climate, which could be interpreted as another form of planetary engineering. By controlling weather patterns and natural disasters, these beings may have been able to influence the development of human civilizations and shape the environment to better suit their needs. Ancient Alien theorists propose that the Watchers' control over Earth's climate could be further evidence of their extraterrestrial origin and their involvement in the development of our planet.

The advanced knowledge provided by the Watchers, according to Ancient Alien theorists, may have also contributed to the construction of monumental structures seen throughout ancient civilizations. The precise construction and alignment of structures such as the pyramids of Egypt, Stonehenge, and the megalithic temples of Malta could be the result of advanced architectural knowledge and technology shared by the Watchers.

These impressive feats of engineering, which remain difficult to explain even with modern technology, could be further evidence of extraterrestrial intervention in human history.

Beyond the physical development of human civilization, some proponents of Ancient Alien Theory suggest that the Watchers may have played a role in the evolution of human consciousness. By imparting advanced knowledge and possibly altering human genetics, the Watchers could have contributed to the development of human intellect, creativity, and spirituality. The rapid advancement of ancient civilizations and the emergence of complex cultural and religious systems might be better understood as the result of extraterrestrial guidance, rather than solely the product of human ingenuity.

The Book of Enoch and the story of the Watchers provide a compelling basis for Ancient Alien theorists to argue for the involvement of extraterrestrial beings in the development of human civilization. The advanced knowledge, genetic manipulation, climate control, architectural advancements, and the evolution of human consciousness attributed to the Watchers offer a fascinating perspective on the possible influence of ancient aliens on our planet's history.

In the Popol Vuh, the god Quetzalcoatl, also known as Kukulkan, is described as a powerful and wise deity who played a significant role in the creation of the Earth and its inhabitants. Quetzalcoatl is often depicted as a serpent or dragon-like creature, which some Ancient Alien theorists suggest could represent an advanced extraterrestrial being. These theorists argue that Quetzalcoatl's actions, such as creating mountains, valleys, and separating the land from the sea, could be interpreted as planetary engineering efforts to shape the Earth's environment and make it more suitable for life.

The ancient Mayan civilization, along with other Mesoamerican cultures, demonstrated remarkable architectural and engineering skills in the construction of their cities and temples. Structures such as the Pyramid of Kukulkan at Chichen Itza and the city of Teotihuacan exhibit sophisticated knowledge of astronomy, geometry, and engineering principles. Some proponents of Ancient Alien Theory argue that these advanced techniques could have been influenced or directly taught by extraterrestrial beings, such as those described in the Popol Vuh. This would further support the idea that ancient aliens were involved in planetary engineering and the development of human civilization.

The Mayan calendar is another example of the advanced knowledge possessed by Mesoamerican civilizations. This intricate and precise system for tracking time is based on detailed astronomical observations, which suggests that the ancient Maya had a deep understanding of celestial mechanics. Some Ancient Alien theorists propose that this advanced knowledge could have been imparted by extraterrestrial beings, who may have also been involved in planetary engineering efforts on Earth. The accurate predictions of celestial events made using the Mayan calendar could be seen as further evidence of an extraterrestrial influence on this ancient civilization.

The Popol Vuh also describes the concept of cyclical destruction and rebirth of the Earth, with multiple creations occurring before the current age. According to the myth, each previous creation was destroyed by the gods, who then reshaped the Earth and created new life. Ancient Alien theorists suggest that this cyclical process could represent a series of terraforming or planetary engineering efforts by extraterrestrial beings, who may have been experimenting with different environmental conditions in an attempt to create the optimal conditions for life to thrive.

The accounts found in the Popol Vuh, along with the advanced knowledge and capabilities attributed to the gods, provide intriguing hints at the possibility of ancient alien involvement in the development of our planet. By studying these myths and the remarkable accomplishments of the ancient Mayan civilization, researchers can gain valuable insights into the potential connections between ancient cultures and extraterrestrial beings. As we continue to explore the cosmos and search for signs of life beyond Earth, the stories and lessons from the Popol Vuh may help guide our understanding of the universe and the potential role of extraterrestrial life in the development of our own world.

The study of ancient texts can provide valuable insights into the knowledge, beliefs, and experiences of past civilizations. While these texts may not always offer definitive proof of ancient alien involvement in planetary engineering, their accounts of advanced technologies, seemingly supernatural beings, and mysterious events can serve as important clues in our quest to understand the relationship between ancient civilizations and extraterrestrial life. By critically examining these texts, researchers can identify patterns and similarities across different cultures and time periods, potentially revealing evidence of extraterrestrial influence on our planet's development.

An interdisciplinary approach to the study of ancient texts can yield valuable insights into the potential connections between extraterrestrial beings and ancient civilizations. By combining the expertise of scholars from various fields, such as archaeology, anthropology, linguistics, and astronomy, researchers can gain a more comprehensive understanding of the historical context and cultural significance of these texts. This collaborative approach can help to uncover the true extent of ancient alien involvement in Earth's history and provide a more accurate picture of the origins of human civilization.

Ancient texts that describe advanced knowledge and technology can serve as a valuable source of inspiration for modern researchers working on planetary engineering and terraforming projects. By analyzing the techniques and strategies outlined in these texts, scientists can potentially uncover innovative solutions to the challenges of transforming other planets into habitable environments. Additionally, the study of ancient texts can help to broaden our understanding of the potential ethical, cultural, and societal implications of planetary engineering, ensuring that future endeavors in this field are conducted responsibly and sustainably.

The accounts found in ancient texts have played a significant role in shaping public perception of Ancient Alien Theory. By highlighting the extraordinary knowledge, technology, and events described in these texts, proponents of the theory can foster a greater appreciation for the potential connections between ancient civilizations and extraterrestrial beings. This increased awareness can, in turn, inspire further research and exploration into the true extent of ancient alien involvement in Earth's history, paving the way for new discoveries and a deeper understanding of our planet's past.

As our understanding of planetary engineering and extraterrestrial life continues to grow, the ancient texts that hint at the possibility of ancient alien involvement in our planet's development will remain an important source of inspiration and insight. By examining these texts through a critical and interdisciplinary lens, researchers can continue to uncover the true extent of ancient alien involvement in Earth's history, shedding light on the origins of human civilization and our place in the cosmos. The ongoing legacy of ancient texts and planetary engineering is a testament to the enduring fascination and curiosity that these accounts inspire, and their potential to inform and shape our understanding of the universe and its inhabitants.

As our technological capabilities advance, we are increasingly able to explore the mysteries of our own planet and the universe beyond. These advancements may lead to the discovery of additional evidence supporting the Ancient Alien Theory, particularly in the context of planetary engineering and terraforming. For example, the development of more sophisticated remote sensing technologies could reveal previously hidden structures or geological features that suggest extraterrestrial involvement in Earth's past.

The ongoing exploration of our solar system and beyond may also provide further insights into the potential existence of ancient extraterrestrial life forms that engaged in planetary engineering. Missions to Mars and other celestial bodies could uncover signs of past or present life, hinting at the possibility that these worlds were once inhabited by intelligent beings with the capability to manipulate their environments. Additionally, the discovery of exoplanets with Earth-like characteristics may further support the notion that advanced alien civilizations have shaped the development of other planets in the cosmos.

The application of artificial intelligence (AI) and machine learning technologies to the study of ancient artifacts, structures, and texts may also contribute to our understanding of the Ancient Alien Theory in the context of planetary engineering. These advanced tools could help researchers detect patterns and anomalies that were previously undetected, potentially revealing new evidence of extraterrestrial intervention in our planet's development. Furthermore, AI could assist in the translation and interpretation of ancient texts, shedding light on the possible connections between ancient civilizations and alien life forms.

As we move forward, interdisciplinary research efforts that combine the expertise of scientists from various fields will play a crucial role in exploring the connections between planetary engineering, terraforming, and Ancient Alien Theory. By fostering collaboration between experts in fields such as geology, climatology, archaeology, and astrobiology, we can gain a more comprehensive understanding of Earth's past and the potential influence of extraterrestrial life on our planet's development.

The exploration of Ancient Alien Theory and its connections to planetary engineering may also have significant implications for our society and our understanding of our place in the cosmos. As we continue to uncover new evidence supporting the idea of ancient extraterrestrial intervention, this knowledge could lead to a paradigm shift in our perception of human history, our origins, and our role in the universe. By acknowledging the potential influence of advanced alien civilizations on our planet's development, we may gain a renewed appreciation for the interconnectedness of all life and the vast possibilities that lie beyond our world.

The future of planetary engineering and Ancient Alien Theory holds immense potential for expanding our understanding of Earth's past and the possible existence of extraterrestrial life forms that have shaped our planet's history. As we continue to develop new technologies, explore the cosmos, and collaborate across disciplines, we may uncover even more evidence that supports the notion of ancient alien intervention in Earth's development, ultimately deepening our appreciation for the complexity and wonder of the universe we inhabit.

The prospect of ancient aliens having played a role in planetary engineering and terraforming on Earth has captivated the minds of many who support the Ancient Alien Theory. As our knowledge of Earth's geological and climatic history expands, so does the potential for uncovering evidence that supports the notion of extraterrestrial involvement in our planet's development. The various anomalies, artifacts, and mythological accounts pointing towards this possibility encourage researchers to delve deeper into the mysteries of our past and reconsider the origins of our world.

The concept of ancient alien involvement in planetary engineering not only impacts our understanding of Earth's history but also has the potential to influence modern scientific advancements. As we explore the possibilities of terraforming other planets and developing new technologies for environmental manipulation, we can look to the alleged ancient alien techniques as a source of inspiration and guidance. This connection between the past and the present may lead to groundbreaking discoveries that propel humanity into a new era of interplanetary exploration and colonization.

The link between ancient aliens and planetary engineering also sheds light on the achievements of ancient civilizations. By considering the possibility that these cultures were influenced by extraterrestrial knowledge and technology, we can reevaluate our understanding of their accomplishments and explore new avenues of research. This perspective may help us unlock long-standing mysteries surrounding the construction of megalithic structures, the development of advanced agricultural systems, and the origins of sophisticated ancient technologies.

As we continue to explore the idea of ancient alien involvement in planetary engineering, we are encouraged to adopt a more open-minded approach to human history. By entertaining the possibility that extraterrestrial beings played a role in shaping our world, we can broaden our perspectives and challenge conventional explanations for the development of life on Earth. This new outlook may lead to exciting discoveries and a deeper understanding of our place in the cosmos.

In conclusion, the potential connection between ancient aliens and planetary engineering remains an intriguing and thought-provoking aspect of Ancient Alien Theory. As we strive to uncover the secrets of our past and unlock the potential for transforming other planets, the legacy of ancient alien planetary engineering will continue to inspire researchers, enthusiasts, and future generations. The quest for answers about our origins and the possible existence of other intelligent life forms that have shaped the course of our planet's history serves as a testament to the enduring fascination with the mysteries of the universe.

Chapter 12

Ancient Texts

The Evidence of Ancient Alien Contact

One of the most compelling aspects of the Anunnaki narrative is their purported role in the creation of humanity. According to Sumerian texts, the Anunnaki created humans to serve as a labor force, assisting them in their endeavors on Earth. This was achieved through a process of genetic engineering, wherein the gods combined their own DNA with that of primitive hominids to create a new, intelligent species.

Proponents of the Ancient Alien Theory argue that this account provides a plausible explanation for the sudden emergence of advanced human civilizations, as well as the significant leap in intellectual and technological capabilities that occurred during this period. They suggest that the Anunnaki's intervention in human evolution may account for the remarkable advancements in areas such as agriculture, architecture, and the arts, which have long puzzled historians and archaeologists.

The Ancient Alien Theory also proposes a connection between the Anunnaki and Nibiru, a hypothetical planet that is said to exist beyond

Neptune in our solar system. According to some interpretations of Sumerian texts, the Anunnaki originated from Nibiru and traveled to Earth in search of precious resources, particularly gold, which they required for various purposes, including the preservation of their planet's atmosphere.

This interpretation of the Anunnaki's motivations for coming to Earth further bolsters the Ancient Alien Theory, as it presents a plausible rationale for their interactions with early human civilizations. Advocates of this theory argue that the Anunnaki's need for gold and other resources would explain their interest in establishing a labor force on Earth, as well as their willingness to share advanced knowledge and technology with humans in order to achieve their goals.

The Sumerian texts also credit the Anunnaki with the construction of many of the ancient world's most impressive and enigmatic structures, such as the Great Pyramid of Giza and the ziggurats of Mesopotamia. These monumental feats of engineering, which continue to astound modern researchers, are often cited as evidence of the Anunnaki's advanced capabilities and their influence on human civilization.

Proponents of the Ancient Alien Theory argue that the construction of these massive structures would have required knowledge and technology beyond the reach of ancient human civilizations, suggesting that extraterrestrial assistance may have been necessary. They point to the precision with which these monuments were built, as well as the incorporation of advanced mathematical and astronomical concepts in their design, as evidence of the Anunnaki's involvement.

The Sumerian texts also describe how the Anunnaki shared their advanced knowledge with select humans, who were chosen to serve as intermediaries between the gods and the general population. These individuals, known as kings or priests, were granted access to the Anunnaki's wisdom and were tasked with disseminating it to the rest of humanity.

The Ancient Alien Theory suggests that this transfer of knowledge played a crucial role in the development of human civilization, providing the foundation for advancements in science, technology, and the arts. By sharing their expertise with early humans, the Anunnaki are believed to have accelerated the progress of human society, shaping the course of history and leaving a lasting legacy that continues to influence our understanding of the world today.

The Rigveda, one of the oldest texts within the collection of the Vedas, contains several hymns dedicated to the gods and their aerial vehicles, the Vimanas. In these hymns, the gods are described as traveling through the sky in their radiant chariots, which are often said to be powered by divine energy or the elements, such as fire or wind. Proponents of the Ancient Alien Theory argue that these descriptions could be interpreted as encounters with extraterrestrial beings who possessed advanced technology, allowing them to traverse the skies in their flying machines.

The Ramayana, another ancient Indian epic, also contains references to Vimanas. In this text, the god Rama is said to have utilized a powerful Vimana known as the "Pushpaka Vimana" during his battles against evil forces. This celestial vehicle is described as having the ability to change its shape, size, and even become invisible at will. Proponents of the Ancient Alien Theory suggest that the Pushpaka Vimana's advanced capabilities could be indicative of extraterrestrial technology, far beyond the understanding of ancient humans.

As previously mentioned, the Mahabharata also contains descriptions of Vimanas, with some passages detailing their use in warfare. The text recounts the story of the hero Arjuna, who is granted the use of a divine Vimana during his battles. This flying machine is said to have been equipped with various weapons, including missiles that could track their targets and destroy them with precision. The advanced nature of these weapons, as well as the aerial capabilities of the Vimanas, has led proponents of the Ancient Alien Theory to argue that these accounts may represent interactions between ancient humans and technologically advanced extraterrestrial beings.

The Vaimanika Shastra, a text purportedly written in the early 20th century but claimed to be a transcription of ancient knowledge, provides detailed information on the construction, operation, and capabilities of Vimanas. The text describes various types of Vimanas, their propulsion systems, and the materials used in their construction. While the authenticity of the Vaimanika Shastra is debated, proponents of the Ancient Alien

Theory argue that it may contain remnants of ancient knowledge that was passed down through generations, potentially originating from extraterrestrial sources.

The presence of Vimanas in ancient Indian texts has significant implications for the Ancient Alien Theory. These accounts of flying machines, with their advanced capabilities and mysterious origins, challenge our understanding of the technological capabilities of ancient Indian civilization. By considering the possibility that Vimanas may be evidence of extraterrestrial contact, we open the door to a greater understanding of the ancient world and the potential influences that have shaped human history.

The Watchers, as described in the Book of Enoch, are said to have descended to Earth and assumed the role of teachers for humanity. They imparted valuable knowledge, including metallurgy, astronomy, medicine, and other advanced sciences. Proponents of the Ancient Alien Theory argue that this transfer of knowledge is indicative of extraterrestrial beings sharing their expertise with ancient civilizations. This idea aligns with the broader theme of gods and divine beings in ancient texts, who are often credited with bestowing wisdom and advanced knowledge upon humans.

Despite their seemingly benevolent intentions, the Watchers' actions were ultimately condemned by the divine order. They were accused of corrupting humanity by revealing forbidden knowledge and engaging in interbreeding with human women, which resulted in the creation of the Nephilim, a race of hybrid beings. Proponents of the Ancient Alien Theory suggest that this narrative could be interpreted as a record of genetic experimentation by extraterrestrial visitors, who sought to create a hybrid species by combining their own genetic material with that of humans.

In the Book of Enoch, the actions of the Watchers led to a series of cataclysmic events, including the Great Flood, which was sent to cleanse the Earth of the corrupt influence of these beings and their hybrid offspring. The Watchers themselves were imprisoned in the depths of the Earth, where they remain to this day. Ancient Alien Theory proponents argue that these events could be an allegorical account of extraterrestrial visitors being forced to leave Earth due to the consequences of their actions or intervention from a higher power.

The story of the Watchers in the Book of Enoch shares similarities with other ancient texts and mythologies, such as the Sumerian accounts of the Anunnaki and the Greek myth of Prometheus, who stole fire from the gods to give to humanity. These parallels have led proponents of the Ancient Alien Theory to suggest that the narratives found in the Book of Enoch and other ancient texts may be different interpretations of the same historical events involving contact between humans and extraterrestrial beings.

The Book of Enoch's account of the Watchers and their imparting of knowledge to humans can be seen as a reflection of the rapid advancements in technology, arts, and sciences that occurred in ancient civilizations. Proponents of the Ancient Alien Theory contend that this leap in human progress was likely influenced by contact with advanced extraterrestrial beings, who shared their knowledge with humanity. This idea challenges conventional explanations for the development of human civilization and invites further investigation into the potential role of extraterrestrial contact in shaping our past.

One aspect that supports the Ancient Alien Theory in relation to the Nazca Lines is the astronomical alignments found within the geoglyphs. Many of the lines and figures appear to correspond with the positions of celestial bodies, such as the sun, moon, and stars. This suggests that the Nazca people had a sophisticated understanding of astronomy and may have been attempting to communicate with extraterrestrial visitors through the use of these alignments. The precision and scale of the Nazca Lines would have required advanced knowledge and technology, further supporting the idea that they were created with the help of otherworldly beings.

The Nazca civilization itself, which thrived from around 100 BCE to 800 CE, presents additional evidence for potential extraterrestrial contact. This ancient culture is known for its complex irrigation systems, advanced pottery techniques, and elaborate textiles. The technological achievements of the Nazca people, coupled with the mysterious nature of the Nazca Lines, have led some proponents of the Ancient Alien Theory to suggest that their civilization was influenced or guided by extraterrestrial beings.

The construction methods employed in creating the Nazca Lines also raise questions about the capabilities of the ancient Nazca people. The lines were made by removing the reddish-brown iron oxide-coated pebbles that cover the surface of the Nazca desert and revealing the light-colored earth underneath. The precision and accuracy with which these massive geoglyphs were created, without the use of modern technology or even a bird's-eye view, point to the possibility of extraterrestrial guidance or intervention. Some Ancient Alien Theory proponents argue that the Nazca people may have used advanced technology or techniques imparted to them by otherworldly visitors.

Another aspect that lends credence to the Ancient Alien Theory is the presence of certain figures and symbols within the Nazca Lines that bear a striking resemblance to modern representations of extraterrestrial beings and spacecraft. For example, some of the humanoid shapes found among the geoglyphs have elongated heads and large, almond-shaped eyes, features commonly associated with the archetypal image of extraterrestrials known as "greys." Additionally, certain geometric patterns found within the Nazca Lines have been compared to modern-day crop circles, which are also often attributed to extraterrestrial activity. These similarities suggest a potential connection between the Nazca Lines and ancient alien visitors.

The enigmatic nature of the Nazca Lines has made them a compelling subject for researchers and proponents of the Ancient Alien Theory. The scale, precision, and astronomical alignments present in the geoglyphs, along with the advanced capabilities of the Nazca civilization, continue to fuel speculation about the possibility of extraterrestrial contact in our ancient past. By considering the Nazca Lines as a potential form of communication with otherworldly beings, we can gain a deeper understanding of the rich and complex history of human civilization and our potential connections to the cosmos.

The Nephilim, as described in the Epic of Gilgamesh and other ancient texts, exhibit traits and characteristics that go beyond the normal capabilities of humans. Their immense size, incredible strength, and superior intellect have led some proponents of the Ancient Alien Theory to argue that these beings were the product of genetic manipulation by advanced extraterrestrial entities. The Nephilim's extraordinary abilities, they contend, could not have arisen through natural evolution alone, and instead point to the influence of otherworldly visitors in shaping human history.

The idea that the Nephilim were the result of hybridization between humans and extraterrestrial beings is further supported by various ancient myths and legends. Stories of gods mating with mortal women and producing offspring with exceptional powers and abilities can be found in the mythologies of numerous cultures across the globe. Proponents of the Ancient Alien Theory suggest that these myths may be based on actual encounters between humans and advanced beings from other worlds, who sought to create a hybrid race that combined the best qualities of both species.

The Nephilim, as depicted in the Epic of Gilgamesh and other ancient texts, were often seen as both revered and feared figures within their societies. Their extraordinary powers and abilities set them apart from ordinary humans, and they were often given positions of authority and leadership. This elevated status, proponents of the Ancient Alien Theory argue, is indicative of the influence and impact that extraterrestrial beings may have had on ancient civilizations. By creating a hybrid race of beings with superior capabilities, these otherworldly visitors could have exerted control over human societies and shaped their development in significant ways.

The Epic of Gilgamesh tells the story of a demigod king's search for immortality, which leads him on a series of adventures and challenges. The quest for immortality is a common theme in many ancient myths and legends, and proponents of the Ancient Alien Theory argue that this recurring motif may be linked to the influence of extraterrestrial beings. The idea that the Nephilim, as the offspring of gods and humans, possessed the secret to eternal life could be seen as an expression of the ancient desire to unlock the secrets of the universe and achieve transcendence through contact with advanced beings from other worlds.

The story of the Nephilim and their extraordinary abilities continues to fascinate and inspire people in modern times. The idea that these beings were the result of genetic manipulation by advanced extraterrestrial entities has become a popular subject in various forms of media, including books, films, and television shows. This enduring fascination with the Nephilim and their potential origins as the offspring of otherworldly visitors is a testament to the power and allure of the Ancient Alien Theory, which invites us to reconsider our understanding of human history and the mysteries of our past.

In addition to the Brahmastra, the Mahabharata describes a range of divine weapons or 'astras,' which were wielded by various characters throughout the epic. These weapons are said to have been bestowed upon the warriors by the gods and possess extraordinary powers that defy conventional understanding of ancient Indian warfare. Some of the notable astras include the Nagastra, which unleashed a torrent of serpents, and the Sammohan Astra, a weapon capable of inducing a deep slumber and confusion among enemies.

Proponents of the Ancient Alien Theory argue that the descriptions of these advanced weapons and their extraordinary capabilities point to the possibility of extraterrestrial intervention. They propose that the divine beings who bestowed these weapons upon the warriors may have, in fact, been technologically advanced extraterrestrial visitors who shared their knowledge and technology with ancient humans.

The Mahabharata also contains references to the aforementioned Vimanas, which are depicted as flying vehicles used by various characters in the epic. These aerial vehicles are described as being capable of incredible feats, such as traveling vast distances at high speeds, hovering, and even engaging in aerial combat.

The detailed accounts of these Vimanas have led proponents of the Ancient Alien Theory to suggest that they could be evidence of advanced extraterrestrial technology that was either observed by or shared with the people of ancient India. The presence of such advanced flying machines in the narrative adds further weight to the argument that ancient humans may have had contact with otherworldly beings possessing advanced technology.

The Mahabharata is not only an epic of warfare and divine intervention but also a treasure trove of knowledge about ancient Indian science, philosophy, and ethics. The text contains vast descriptions of cosmology, astronomy, medicine, and mathematics, which were far ahead of their time.

For instance, the concept of time dilation – the idea that time flows at different rates for different observers – is discussed in the Mahabharata, a concept that was not formalized in modern physics until the 20th century with the advent of Einstein's theory of relativity.

Proponents of the Ancient Alien Theory argue that the advanced knowledge contained within the Mahabharata could be further evidence of extraterrestrial influence on ancient Indian civilization. They propose that the sophisticated concepts and technologies described in the text may have been transmitted by otherworldly beings who sought to share their knowledge with humanity.

The Mahabharata serves as a compelling example of how ancient texts can be reexamined through the lens of the Ancient Alien Theory. The epic's detailed accounts of advanced weaponry, flying machines, and sophisticated scientific concepts challenge conventional understanding of ancient Indian civilization and invite speculation about the potential involvement of extraterrestrial beings in human history. While this interpretation may not be universally accepted, it nevertheless encourages us to reevaluate our understanding of the past and consider the possibility that the knowledge and technologies of ancient civilizations may have originated from sources beyond our own planet.

The Popol Vuh provides a detailed account of the genesis of humanity, describing the various attempts of the gods to create beings capable of worshiping and communicating with them. The gods initially fashioned beings made of mud, but these creations proved to be unsatisfactory, as they lacked intelligence and soon crumbled. Next, the gods attempted to create humans from wood; however, these beings lacked emotion and consciousness, leading the gods to destroy them in a great flood.

Finally, the gods discovered maize, a staple crop in Mesoamerican culture, and used maize dough to create the first humans. These new beings possessed intelligence, emotions, and the ability to communicate with their creators. The inclusion of maize, a critical source of sustenance in ancient Mesoamerica, highlights its cultural significance and its role as a symbol of life.

Proponents of the Ancient Alien Theory argue that the process of trial and error described in the Popol Vuh could be a metaphor for advanced genetic engineering performed by extraterrestrial beings. The gods' various attempts to create humans might represent the experimentation and refinement of genetic manipulation techniques to develop a sentient, intelligent species. The eventual success in creating humans from maize dough could symbolize the incorporation of Earth-based genetic material with extraterrestrial elements, resulting in the birth of modern humans.

The Popol Vuh also recounts the adventures of the Hero Twins, Hunahpu and Xbalanque, who are sent by the gods to defeat the lords of the underworld. These heroes display extraordinary abilities, which could be interpreted as evidence of advanced technology or knowledge bestowed upon them by their divine creators. Their feats are reminiscent of other ancient myths in which heroes perform extraordinary deeds with the help of otherworldly beings or technology.

The Popol Vuh's account of humanity's creation provides an intriguing perspective on the origins of our species and the potential influence of extraterrestrial beings on our development. As one of the many ancient texts that describe divine intervention in the creation of human civilization, the Popol Vuh invites us to consider the possibility that the gods of mythology may, in fact, be ancient aliens who played a crucial role in shaping the course of human history.

The Dogon people's understanding of astronomy extends far beyond their knowledge of the Sirius star system. They possess a detailed awareness of celestial bodies, their movements, and their relationships to one another. For example, they are aware of the elliptical orbit of Sirius B around Sirius A, a fact that was only confirmed by modern science in the late 20th century. Additionally, the Dogon possess knowledge about the planets in our solar system, their respective orbits, and the fact that the Earth revolves around the Sun. This extensive comprehension of astronomy is remarkable, especially considering the lack of advanced telescopes or other tools that would have been required to observe these celestial phenomena.

Central to the Dogon's cosmology is the story of the Nommo, a group of amphibious beings who are said to have descended from the stars to create the Earth and its inhabitants. The Dogon believe that the Nommo came from the Sirius star system, specifically from Sirius B. According to their mythology, the Nommo created the first human beings and imparted them with the knowledge of agriculture, social organization, and other aspects of civilization.

Proponents of the Ancient Alien Theory argue that the Dogon's intricate creation myth, which includes specific details about the Nommo's origin in the Sirius system, could be evidence of extraterrestrial contact. They suggest that the Nommo may have been advanced beings from another planet who visited Earth in the distant past, providing the Dogon with knowledge that would have been otherwise unattainable.

The Dogon's cultural practices further support the possibility of ancient alien contact. Their rituals and ceremonies often revolve around the stars and celestial events, with a particular emphasis on the Sirius system. For instance, the Dogon celebrate a festival called Sigui, which occurs every 60 years and corresponds to the orbital period of Sirius B. This remarkable alignment suggests that the Dogon possess an intimate understanding of the Sirius system's movements, which is difficult to explain without the intervention of a more advanced civilization.

The Dogon people's art and symbolism also reflect their deep connection to the stars and their understanding of the cosmos. Their artwork often features complex geometric patterns and representations of celestial bodies, as well as images of the Nommo and other mythological beings. The precision and intricacy of the Dogon's art suggest a level of sophistication and knowledge that is difficult to reconcile with the limited technological capabilities of their time.

The Dogon people's advanced astronomical knowledge, their intricate creation myth involving the Nommo from the Sirius star system, and their cultural practices and art all point towards the possibility of ancient alien contact. While it is difficult to definitively prove the extraterrestrial origins of the Dogon's knowledge, the striking correlations between their beliefs and modern scientific understanding of the cosmos invite us to consider the potential influence of otherworldly beings on this ancient civilization.

Zoroastrian cosmology, with its intricate descriptions of the celestial realm, has fueled speculations about possible extraterrestrial connections.

The texts describe various celestial beings, such as the Amesha Spentas, who serve as divine representatives of Ahura Mazda's attributes. The Amesha Spentas are said to govern different aspects of the natural world, including the elements, animals, and human virtues.

Proponents of the Ancient Alien Theory argue that the detailed accounts of these celestial beings could be inspired by encounters with extraterrestrial entities. The advanced knowledge associated with these beings, as well as their role in maintaining cosmic order, might be indicative of a more technologically advanced civilization sharing its wisdom with ancient humans.

Zoroastrian rituals, such as the Yasna, a central rite involving the consecration of the sacred drink Haoma, have also been cited as potential evidence of extraterrestrial influence. The Haoma, a plant with mystical properties, is said to grant strength, health, and even immortality to those who consume it.

Ancient Alien Theory proponents suggest that the rituals surrounding the Haoma plant could be a result of ancient humans attempting to replicate advanced extraterrestrial technologies or substances. The plant's extraordinary properties might be a reflection of otherworldly knowledge or technology introduced to ancient Persian society by extraterrestrial visitors.

The Zoroastrian texts also contain prophecies about the end of the world and the eventual triumph of good over evil, known as the Frashokereti. This event is said to be preceded by a series of cosmic battles between the forces of Ahura Mazda and Angra Mainyu, culminating in the restoration of the world to its original, perfect state.

Some proponents of the Ancient Alien Theory argue that these prophecies could be interpreted as a coded message from extraterrestrial beings, who may have foreseen or even planned for future events that would impact the Earth. The concept of a cosmic battle between good and evil might represent an extraterrestrial intervention in human affairs, with the ultimate goal of guiding humanity toward a better future.

The Zoroastrian texts and their intricate cosmology, rituals, and prophecies offer a rich source of potential evidence for the Ancient Alien Theory. The dualistic worldview, advanced philosophical concepts, and detailed accounts of celestial beings and events might be interpreted as the result of contact between ancient humans and extraterrestrial visitors. While these connections remain speculative, they invite us to consider the possibility that our ancestors' religious beliefs and practices were influenced by otherworldly encounters, which continue to shape our understanding of the universe and our place within it.

One of the most intriguing aspects of the Voynich Manuscript is the unknown language in which it is written. Despite extensive efforts by linguists, cryptographers, and historians, the script has yet to be deciphered or even identified. Proponents of the Ancient Alien Theory argue that the complexity and uniqueness of this language may be the result of an extraterrestrial origin. The idea that the language could be a form of communication used by otherworldly beings visiting Earth in the past raises the possibility that the manuscript contains knowledge and information that humans have yet to comprehend.

The Voynich Manuscript contains numerous illustrations of celestial bodies, plants, and animals, many of which do not resemble any known species on Earth. These peculiar images have led some proponents of the Ancient Alien Theory to suggest that the manuscript could contain depictions of extraterrestrial flora and fauna or even portray aspects of other planets and their ecosystems. The presence of such illustrations in the manuscript could be interpreted as evidence of ancient contact with extraterrestrial beings who shared their knowledge of the cosmos and its diverse life forms.

Another aspect of the Voynich Manuscript that has captured the interest of Ancient Alien Theory proponents is the possibility that it contains advanced concepts and technologies beyond the understanding of 15th-century human civilization. The cryptic nature of the text and the enigmatic illustrations might be an attempt to convey information about scientific principles, engineering techniques, or even medical procedures that were unfamiliar to the manuscript's creator. If this is the case, it raises the question of how such advanced knowledge was obtained and whether it could have been the result of extraterrestrial contact.

The origins of the Voynich Manuscript are shrouded in mystery, with no clear indication of who created it or for what purpose. This lack of information has led to numerous theories and speculations, including the possibility that the manuscript was created by an extraterrestrial visitor or someone who had contact with otherworldly beings. The idea that the manuscript might be a product of ancient alien interaction adds another layer of intrigue to the ongoing efforts to decipher its contents and understand its purpose.

The enduring enigma of the Voynich Manuscript, despite decades of research and countless attempts to unlock its secrets, has fueled the interest of Ancient Alien Theory proponents. The manuscript's resistance to interpretation and the many unanswered questions surrounding its origins, language, and illustrations could be seen as evidence of a connection to otherworldly beings and their advanced knowledge. As the search for answers continues, the Voynich Manuscript remains a tantalizing and mysterious artifact that invites us to consider the possibility of ancient contact with extraterrestrial civilizations and the potential impact of such interactions on human history.

The origins of the Emerald Tablet are shrouded in mystery, with various accounts attributing its creation to different historical figures, including Hermes Trismegistus, a legendary philosopher and alchemist. Hermes Trismegistus, often considered a syncretic figure combining the Greek god Hermes and the Egyptian god Thoth, is said to have been an intermediary between humans and the gods. This connection has led proponents of the Ancient Alien Theory to propose that Hermes Trismegistus could have been an extraterrestrial being or a human who had contact with otherworldly visitors.

The Emerald Tablet is believed to contain the secrets of the universe and the process of transmutation, a cornerstone of alchemical practice. Transmutation, the process of turning base metals into gold or other precious substances, has long been seen as a metaphor for spiritual transformation and enlightenment. The advanced knowledge and principles found in the Emerald Tablet have led proponents of the Ancient Alien Theory to speculate that its teachings could have originated from extraterrestrial beings with a profound understanding of the natural world and the secrets of the cosmos.

Throughout history, the Emerald Tablet has had a profound impact on the development of alchemy, Hermetic philosophy, and the esoteric traditions of various cultures. Some of the most influential thinkers and scholars, such as Isaac Newton, Paracelsus, and Roger Bacon, have studied and commented on the Emerald Tablet. Proponents of the Ancient Alien Theory argue that the widespread influence of this mysterious text on human civilization is indicative of its otherworldly origins, as it has transcended cultural and temporal boundaries to shape the course of human intellectual and spiritual development.

The idea that the wisdom contained within the Emerald Tablet was transmitted by otherworldly beings reflects a broader pattern of ancient texts that describe advanced concepts and technologies beyond the capabilities of the cultures that produced them. The enduring mystery surrounding the origins and contents of the Emerald Tablet has fueled speculation that it may be a repository of lost knowledge from an advanced extraterrestrial civilization.

In this context, the Emerald Tablet serves as a testament to the potential influence of ancient alien contact on the development of human civilization, particularly in the realms of science, philosophy, and spirituality. As we continue to explore the depths of our own understanding and the mysteries of the universe, the legacy of the Emerald Tablet and its possible connection to extraterrestrial beings challenges us to consider the potential impact of otherworldly wisdom on the course of human history.

The Abydos Helicopter, as it has come to be known, was first brought to public attention in the 1990s. These intriguing hieroglyphs are located on a beam in the Temple of Seti I, a structure dedicated to the pharaoh Seti I and his son, Ramesses II. The discovery of these carvings has generated significant debate among scholars and enthusiasts alike, with proponents of the Ancient Alien Theory arguing that the images are evidence of ancient Egyptians encountering advanced technology that could only have been introduced by extraterrestrial beings.

Upon closer examination, the Abydos Helicopter appears to consist of several modern vehicles, including a helicopter with a distinct rotor, a submarine with a streamlined hull and fins, and an aircraft with a clear tail and wings. These images are strikingly detailed and closely resemble 20th-century vehicles, which is particularly remarkable given that the Temple of Seti I dates back to around 1280 BCE.

Mainstream Egyptologists argue that the Abydos Helicopter is the result of overlapping carvings created during different time periods. They suggest that the original hieroglyphs were later re-carved and modified, resulting in the seemingly modern images. However, proponents of the Ancient Alien Theory maintain that the precision and resemblance of these carvings to contemporary vehicles cannot be dismissed as mere coincidence or the result of overlapping carvings.

The Abydos Helicopter is just one of many examples of potential extraterrestrial influence in ancient Egypt. Ancient Alien Theory proponents argue that the advanced knowledge and technology demonstrated by the ancient Egyptians, including the construction of the pyramids, the use of complex mathematical concepts, and the development of sophisticated medical practices, could be attributed to contact with extraterrestrial beings. These beings may have shared their advanced technology and knowledge with the Egyptians, resulting in the impressive accomplishments of their civilization.

The discovery of the Abydos Helicopter and its potential connection to extraterrestrial technology has far-reaching implications for our understanding of human history. If these hieroglyphs are indeed evidence of ancient alien contact, it challenges conventional explanations of the development of technology and raises questions about the true extent of extraterrestrial influence on Earth. This possibility opens up new avenues of exploration, both in terms of understanding our past and in considering the potential existence of other civilizations in the cosmos.

In conclusion, the Ancient Alien Theory offers a unique perspective on human history, one that suggests the intervention of extraterrestrial beings may have significantly shaped the development of our civilization. By examining ancient texts and artifacts through this lens, we uncover intriguing connections and possibilities that challenge conventional explanations and invite further inquiry.

The stories of advanced knowledge, technologies, and seemingly divine encounters found in ancient texts from various cultures around the world suggest that there might be more to our past than meets the eye. The enduring legacy of these accounts prompts us to consider the possibility that extraterrestrial beings may have played a role in shaping our history, either by direct intervention or by inspiring ancient humans with their advanced capabilities.

Furthermore, the Ancient Alien Theory encourages a reevaluation of historical narratives and the way we understand human progress. By entertaining the possibility of extraterrestrial contact, we open ourselves up to a more expansive view of history, one that acknowledges the vastness of the universe and the potential for life beyond our planet. This perspective also fosters a sense of humility, as it reminds us that we may not be the only intelligent beings in the cosmos, and that our understanding of the universe is still in its infancy.

Ultimately, the Ancient Alien Theory serves as a catalyst for inquiry and exploration, inviting us to delve deeper into the mysteries of our past and consider the potential impact of otherworldly beings on human civilization. As we continue to uncover new evidence and expand our knowledge of the cosmos, the Ancient Alien Theory will undoubtedly remain an influential and thought-provoking force in our ongoing quest to understand our origins and our place in the universe.

Chapter 13

Technological Evidence

Clues of Alien Involvement in Our Past

Ancient Alien Theory proponents argue that there are numerous examples of advanced technology and knowledge present in ancient artifacts and structures, which are difficult to explain using conventional historical accounts. They claim that these examples indicate the involvement of extraterrestrial beings in the development of ancient civilizations. This chapter will explore the most compelling pieces of technological evidence that support the idea of ancient alien interaction, analyzing the construction methods, materials, and the implications of these discoveries.

The Great Pyramid of Giza stands as one of the most enigmatic and impressive architectural feats in human history. Despite extensive research and analysis, the precise methods and techniques employed in its construction remain shrouded in mystery. The precision and scale of the Great Pyramid have led many researchers to consider the possibility that its builders received assistance from a more technologically advanced civilization, perhaps even extraterrestrial in origin.

One of the most striking features of the Great Pyramid is its remarkable alignment with the cardinal points. The base of the pyramid is a near-perfect square, with each side aligned to within a fraction of a degree of true north, south, east, and west. This level of accuracy would be challenging to achieve even with modern tools and equipment, raising questions about how the ancient Egyptians could have accomplished such a feat without access to advanced technology or knowledge.

The construction of the Great Pyramid required the movement and placement of massive stone blocks, some weighing up to 70 tons. These blocks were quarried from sites hundreds of kilometers away, and their transportation to the pyramid site would have been an immense logistical challenge. The methods by which the ancient Egyptians were able to move and position these enormous stones with such precision remain unknown, further fueling speculation about potential extraterrestrial assistance.

The Great Pyramid's internal structure is equally as impressive as its exterior. The pyramid contains multiple chambers and passageways, including the King's Chamber, which features a corbelled ceiling designed to distribute the weight of the stone blocks above it. This innovative engineering solution, along with other unique construction techniques found within the pyramid, demonstrates an advanced understanding of mathematics, geometry, and structural engineering that appears to be beyond the capabilities of the ancient Egyptians.

The Great Pyramid's purpose has long been debated, with theories ranging from a tomb for the pharaoh Khufu to a monument commemorating an astronomical event. Ancient Alien theorists suggest that the pyramid may have served a more profound purpose, possibly as a beacon or energy source for extraterrestrial visitors. The intricate design and construction of the Great Pyramid, combined with its alignment to celestial bodies, lend credence to the idea that it may have been built with the guidance of an advanced civilization.

The Great Pyramid of Giza stands as a testament to the remarkable capabilities of its builders and raises questions about the possible involvement of extraterrestrial beings in its construction. The precision, scale, and engineering techniques employed in the pyramid defy conventional explanations, leading many to consider the possibility that its creation was guided by advanced knowledge from beyond our planet.

Puma Punku, an archeological site in Bolivia, continues to mystify researchers with its advanced stone-cutting techniques and the complexity of its design. The ruins are made up of intricately carved stone blocks, displaying an impressive level of precision in their angles and surfaces. The complex interlocking designs found on these blocks would have necessitated a deep understanding of geometry and engineering that far exceeds what was believed to be possible for the ancient cultures of the region.

The age of Puma Punku is estimated to be at least 2,000 years old, making it significantly older than the Inca civilization. This presents a conundrum for researchers, as there is no known culture from that time period in the region that could have possessed the advanced knowledge and skills required to construct such a sophisticated site. The unexplained origins of Puma Punku lead some proponents of the Ancient Alien Theory to suggest that extraterrestrial beings may have played a role in its construction, providing guidance or direct assistance to the builders.

One of the most compelling aspects of Puma Punku is the remarkable stonework found throughout the site. The stones used in the construction are not only precisely shaped and polished but are also made of incredibly hard materials such as diorite and andesite. These materials would have required the use of advanced tools and techniques to cut and shape, yet no evidence of such tools has been discovered at the site. Ancient Alien theorists argue that this could be a sign of extraterrestrial involvement, as the level of craftsmanship displayed at Puma Punku seems to be far beyond the capabilities of the known civilizations of the time.

Another intriguing aspect of Puma Punku is the transportation and assembly of the massive stone blocks used in its construction. Some of these stones weigh up to 100 tons and are believed to have been quarried from sites several kilometers away. The challenges involved in moving and positioning these enormous blocks are immense, even by today's standards. It is difficult to imagine how an ancient culture could have achieved this feat without the use of advanced technology or assistance from an outside source. Ancient Alien proponents propose that extraterrestrial beings may have provided the necessary technology or knowledge to facilitate the transportation and assembly of these massive stones.

Puma Punku stands as a testament to the remarkable engineering and construction capabilities of the ancient world. The unexplained origin, advanced stonework, and the challenges involved in its construction lend credence to the idea that extraterrestrial beings may have played a role in its creation. The site serves as a compelling piece of evidence in favor of Ancient Alien Theory, suggesting that our understanding of human history may need to be reevaluated to account for the possibility of extraterrestrial involvement in the development of ancient civilizations.

The Antikythera Mechanism is an intriguing and enigmatic artifact that continues to captivate researchers and Ancient Alien Theory proponents alike. The intricate nature of its design, combined with the precision and complexity of its gears, suggests that it was the work of a civilization with advanced knowledge of astronomy, engineering, and mathematics.

The mechanism is comprised of a series of over 30 bronze gears, housed in a wooden frame, that work together to simulate the movements of celestial bodies. The level of precision and accuracy exhibited in the gear ratios and interlocking components is astounding, particularly considering the era in which it was created. The Antikythera Mechanism's design is so complex that it rivals the work of mechanical engineers and clockmakers in the 18th century, a full 2,000 years after it was supposedly built.

The Antikythera Mechanism was capable of predicting solar and lunar eclipses, as well as tracking the movements of the Sun, Moon, and several planets. This degree of astronomical knowledge was far ahead of its time and suggests that the device's creators had a deep understanding of the mechanics of the cosmos. Such knowledge seems unlikely to have been attained by the ancient Greeks alone, leading some to propose that they may have been aided by extraterrestrial beings with advanced astronomical knowledge.

The existence of the Antikythera Mechanism raises important questions about the technological capabilities of ancient civilizations. If such an advanced device could be created more than 2,000 years ago, it suggests that our understanding of ancient technological prowess is significantly underestimated. This device's complexity and sophistication imply that the creators had access to a level of knowledge that might have been shared or guided by an extraterrestrial civilization.

Proponents of the Ancient Alien Theory argue that the Antikythera Mechanism is a prime example of the potential influence of extraterrestrial beings on ancient human civilizations. The remarkable engineering and astronomical knowledge exhibited in this artifact are difficult to reconcile with the conventional historical narrative. As a result, the Antikythera Mechanism serves as a compelling piece of evidence that points to the possibility of ancient alien contact, revealing the breadth and depth of their impact on human history.

The discovery of the Baghdad Battery in Iraq has sparked significant interest in the field of ancient technology, as it presents a compelling case for the existence of ancient electrical devices. The artifacts consist of terracotta jars, each housing an iron rod encased in a copper cylinder. These components suggest a rudimentary galvanic cell, which, when filled with an electrolytic solution such as vinegar or lemon juice, could generate a small electrical current.

The presence of an electrical device in ancient Mesopotamia has far-reaching implications, as it suggests a level of technological understanding that far surpasses what historians previously believed to be possible in that era. The ability to harness electricity opens the door to a multitude of potential applications, ranging from simple electroplating to more complex devices that could have been used for communication, lighting, or even medical purposes.

The unexplained nature of the Baghdad Battery lends credence to the Ancient Alien Theory, as it raises questions about the origin of this advanced knowledge. Proponents of the theory argue that extraterrestrial beings may have shared their expertise with ancient civilizations, providing them with the means to develop and utilize electrical devices. This exchange of knowledge could have had a profound impact on the development of ancient Mesopotamian society, influencing everything from their art and architecture to their religious beliefs and practices.

The Baghdad Battery is not the only instance of potential ancient electrical devices. Other artifacts and accounts from various cultures around the world suggest that the knowledge of electricity may have been more widespread than previously thought. For example, Egyptian hieroglyphs depict what some believe to be scenes of electrical lighting, while ancient Greek texts mention the use of electric fish for pain relief. These examples, when considered alongside the Baghdad Battery, strengthen the argument for extraterrestrial influence in the development and dissemination of advanced technological knowledge throughout the ancient world.

The Vimanas, as described in the ancient Indian texts, are remarkable flying machines with capabilities that seem to surpass the technology of the time. The texts portray these vehicles as chariots of the gods, often large enough to transport several passengers and equipped with advanced weaponry. The Vedas and other texts, such as the Mahabharata and the Ramayana, provide detailed accounts of these machines' abilities, including their capacity to become invisible, change their shape, and emit beams of light. These descriptions evoke images of modern-day aircraft, spacecraft, and even UFOs, raising questions about the origins of this knowledge.

Ancient Indian texts also discuss the technology that powered the Vimanas, mentioning mercury vortex engines, solar power, and other energy sources that seem far ahead of the civilizations of the time. The propulsion systems described in these texts are strikingly similar to some theories of advanced spacecraft propulsion being researched today, such as ion propulsion and nuclear fusion drives. This similarity between ancient Vimana technology and contemporary space travel concepts has led some researchers to propose that the ancient Indians might have had access to extraterrestrial knowledge or that they were directly inspired by encounters with advanced alien technology.

The Vedas and other ancient Indian texts also demonstrate an extensive understanding of aeronautical concepts, discussing principles such as lift, drag, and propulsion, which are essential to modern aviation. These texts even describe different types of Vimanas, each with its unique design and purpose, similar to the variety of aircraft we see today. This advanced aeronautical knowledge, combined with the detailed descriptions of Vimana technology, further supports the idea that ancient Indians may have had access to information beyond their time, potentially through contact with extraterrestrial beings.

The detailed accounts of Vimanas in ancient Indian texts provide compelling evidence for the Ancient Alien Theory. The technology and capabilities described in these texts are difficult to explain using conventional historical accounts, suggesting that the ancient Indians might have been influenced by extraterrestrial visitors or had access to otherworldly knowledge. By examining the Vimanas and their associated technology, we can gain valuable insights into the possible interactions between ancient civilizations and advanced extraterrestrial beings, paving the way for a better understanding of our own history and our place in the cosmos.

The Sumerian civilization, which emerged around 4500 BCE in Mesopotamia, left behind a wealth of texts, including cuneiform tablets that mention the Anunnaki. These texts describe the Anunnaki as a race of gods who played a significant role in the development of Sumerian society, imparting knowledge in various fields, such as agriculture, mathematics, and astronomy. These beings were also credited with the creation of humanity, suggesting a direct link between the Anunnaki and the advancements of early human civilization.

The accounts of the Anunnaki's involvement in the development of the Sumerian civilization suggest that they possessed advanced knowledge and capabilities that were beyond the understanding of the Sumerians themselves. This includes the Anunnaki's purported ability to create and manipulate life, as well as their mastery of various forms of advanced technology. The Sumerians credited the Anunnaki with teaching them how to build and maintain complex canal systems, which enabled the Sumerians to develop agriculture and sustain their civilization in the arid environment of Mesopotamia.

Ancient Alien theorists propose that the Anunnaki were not mythical deities but rather extraterrestrial beings who visited Earth and interacted with early human civilizations. This perspective is based on the advanced knowledge and technology attributed to the Anunnaki in Sumerian texts, which seem to be beyond the capabilities of humans at that time. Theorists also point to the depictions of the Anunnaki in Sumerian art, where they are often portrayed as humanoid figures with distinctive features such as elongated skulls or bird-like wings, suggesting a possible non-human origin.

The idea that the Anunnaki were extraterrestrial visitors who shared their knowledge and technology with the Sumerians has been embraced by various researchers and enthusiasts of the Ancient Alien Theory. This interpretation challenges the conventional understanding of human history and highlights the potential for advanced civilizations to have existed in the distant past, possibly with extraterrestrial assistance. By considering the possibility that the Anunnaki were alien beings, we gain a fresh perspective on the origins of human civilization and our potential connections to other intelligent life forms in the cosmos.

Trepanation, an ancient surgical procedure involving the drilling or cutting of a hole in the skull, has been documented in various cultures across the globe, from ancient Europe to South America. The survival rates for individuals who underwent trepanation in antiquity are surprisingly high, with some estimates suggesting that up to 70% of patients survived the procedure. This success rate is remarkable, given the rudimentary tools and lack of modern medical knowledge available during these times. The survival rates, along with the widespread practice of trepanation, raise questions about the source of the advanced medical knowledge required to perform such a delicate operation.

The techniques and tools used in trepanation varied among different ancient cultures, but all methods required a high level of skill and precision. Some cultures used sharpened stones or metal implements to create a circular groove around the area to be removed, while others used drilling techniques with rudimentary tools. The fact that these ancient civilizations could perform such complex procedures with limited technology suggests a level of understanding and expertise that seems out of place in the historical context. Ancient Alien theorists argue that this expertise could be the result of guidance or direct intervention from extraterrestrial beings.

The advanced nature of trepanation and the prevalence of the practice across ancient cultures with seemingly no direct contact have led proponents of the Ancient Alien Theory to consider the possibility of extraterrestrial involvement. They argue that the level of skill required to perform trepanation and the high survival rates suggest that ancient surgeons may have been taught or influenced by beings with advanced medical knowledge. These extraterrestrial beings could have shared their expertise with different ancient civilizations, enabling them to develop advanced surgical techniques that seem out of place in the context of their time.

The practice of trepanation provides a compelling piece of evidence for the Ancient Alien Theory, demonstrating the existence of advanced medical knowledge in ancient civilizations that is difficult to explain using conventional historical accounts. The widespread nature of trepanation, along with the skill and precision required to perform the procedure, supports the idea that ancient surgeons may have benefited from extraterrestrial guidance or intervention. By considering the potential connection between trepanation and ancient alien involvement, we can further explore the possibility of extraterrestrial influence in human history and better understand the development of medical knowledge throughout the ages.

The Ica Stones, a collection of engraved stones discovered near the city of Ica in Peru, have been a source of fascination and controversy among researchers and Ancient Alien theorists alike. These stones, which number in the thousands, feature intricate carvings that appear to depict advanced medical procedures, astronomical observations, and other scenes indicative of a highly sophisticated civilization.

The Ica Stones' illustrations of complex medical procedures, such as heart transplants and caesarean sections, are particularly intriguing. These carvings suggest a level of surgical knowledge and skill that far surpasses what would be expected of ancient civilizations. Ancient Alien theorists argue that this advanced medical knowledge could have been provided by extraterrestrial beings who visited Earth and shared their expertise with ancient humans. This perspective offers an explanation for the seemingly impossible medical feats depicted on the Ica Stones.

In addition to the depiction of advanced medical procedures, the Ica Stones also feature carvings that suggest the use of advanced technology for astronomical observations. These carvings include images of what appear to be telescopes, as well as detailed representations of celestial bodies and constellations. The level of astronomical knowledge and technology implied by these carvings is difficult to reconcile with the historical context in which they were created. Ancient Alien theorists propose that this knowledge could be the result of extraterrestrial beings sharing their advanced technology and understanding of the cosmos with ancient humans.

While the authenticity of the Ica Stones has been the subject of debate, the possibility that they represent genuine evidence of advanced medical and technological knowledge in ancient civilizations cannot be easily dismissed. If the Ica Stones are indeed authentic, they could provide valuable clues about the existence of ancient aliens and their potential influence on human history. By considering the possibility of extraterrestrial involvement in the development of ancient civilizations, we may gain a deeper understanding of our past and a greater appreciation for the mysteries that continue to challenge our conventional views of history.

The Nazca Lines, situated in the Nazca Desert in southern Peru, are a collection of vast geoglyphs representing an assortment of animals, plants, and geometric shapes. These impressive drawings, visible in their entirety only from the air, have prompted researchers to ponder their intended purpose and the technology employed in their creation. Proponents of the Ancient Alien Theory posit that the Nazca Lines may have served as markers or landing strips for extraterrestrial spacecraft.

The extraordinary precision and scale of the Nazca Lines suggest that their creators possessed advanced knowledge of geometry and cartography. The lines, which stretch for miles and maintain their proportions and straightness, indicate that the Nazca people could have been aided by a more technologically advanced civilization or guided by extraterrestrial beings with aerial perspectives. The level of accuracy and consistency displayed in these geoglyphs is difficult to explain using the conventional understanding of the Nazca culture's technological capabilities.

The fact that the Nazca Lines can only be fully appreciated from an aerial viewpoint raises questions about their intended purpose. Ancient Alien theorists argue that this aerial perspective suggests the involvement of beings with the ability to view the geoglyphs from above – either from a high altitude or space. This idea is supported by the fact that the Nazca Lines were created over a thousand years before the invention of human flight, making it improbable that the Nazca people would have created these massive drawings without the knowledge or assistance of extraterrestrial visitors.

Some researchers have found connections between the Nazca Lines' design and astronomical alignments, indicating that the geoglyphs may have served a celestial purpose. This advanced understanding of astronomy and its integration into the Nazca Lines further supports the idea that the Nazca people were aided by extraterrestrial beings with a sophisticated knowledge of the cosmos. The incorporation of astronomical alignments into the geoglyphs' design hints at a possible communication between the Nazca culture and visitors from the stars.

The Nazca Lines bear striking similarities to other ancient sites worldwide, such as the Uffington White Horse in England and the Marree Man in Australia, both of which are large-scale geoglyphs visible only from the air. These similarities suggest a shared knowledge or influence among ancient cultures, which could be attributed to contact with extraterrestrial beings. The existence of these parallel sites across different continents and time periods lends credence to the Ancient Alien Theory, suggesting a global pattern of alien involvement in human history.

The Ubaid archaeological site in Iraq has revealed a series of ancient artifacts, including statues that depict humanoid figures with elongated skulls, large eyes, and seemingly wearing helmets. These "helmeted" figures bear a striking resemblance to modern descriptions of extraterrestrial beings, leading Ancient Alien proponents to argue that these artifacts may represent evidence of ancient encounters with otherworldly visitors. The advanced technology and knowledge that these beings could have shared with ancient civilizations might have influenced the development of art and culture in the region.

The prehistoric cave paintings in Tassili n'Ajjer, Algeria, have fascinated researchers due to their depictions of humanoid figures wearing what some interpret as spacesuits or advanced protective gear. These figures are often accompanied by strange objects that resemble modern UFOs or advanced aerial vehicles. Ancient Alien theorists suggest that these cave paintings may be a visual record of ancient encounters with extraterrestrial beings and their advanced technology, possibly influencing the culture and beliefs of the people who created these artworks.

Several Renaissance paintings also feature objects that resemble modern UFOs, including "The Annunciation with Saint Emidius" by Carlo Crivelli and "The Madonna with Saint Giovannino" by Domenico Ghirlandaio. In these paintings, peculiar objects in the sky emit beams of light, drawing parallels with contemporary UFO sightings. Ancient Alien proponents argue that these artistic depictions of otherworldly objects indicate a long history of human interaction with extraterrestrial beings and their technology, possibly influencing the development of art and culture during the Renaissance period.

A 16th-century woodcut from Nuremberg, Germany, documents an unusual celestial event that took place in 1561. The woodcut depicts a chaotic scene of various objects in the sky, including spheres, cylinders, and crosses, engaged in what appears to be a battle or interaction of some sort. Ancient Alien theorists suggest that this event may have been an encounter between extraterrestrial spacecraft or an example of advanced technology beyond the comprehension of people at the time. The enduring mystery surrounding this event and its depiction in art highlights the potential influence of otherworldly encounters on human creativity and culture throughout history.

Carvings and sculptures found in various ancient sites worldwide have been interpreted by Ancient Alien proponents as potential evidence of ancient encounters with extraterrestrial beings. Examples include the Mayan carving known as the "Palenque Astronaut," which appears to depict a figure seated in a complex machine, and the Dogu figurines from ancient Japan, which resemble modern depictions of space travelers in protective suits. These artistic representations of what appear to be advanced technology and otherworldly beings support the notion that ancient civilizations may have had contact with extraterrestrial visitors, who may have shared their knowledge and influenced the development of art and culture.

The various examples of advanced technology and knowledge presented in this chapter highlight significant gaps in our understanding of the development of ancient civilizations. The intricate construction techniques, precise astronomical alignments, and complex devices, such as the Antikythera Mechanism, all suggest that these societies may have had access to knowledge and tools that were beyond their time. By embracing the Ancient Alien Theory, we can begin to explore the possibility that these technological marvels were the result of extraterrestrial intervention, providing a fresh perspective on the growth and evolution of human knowledge.

The idea of ancient alien involvement in our past encourages us to question the limits of human potential and ingenuity. If these advanced technologies were the result of collaboration between our ancestors and extraterrestrial beings, it implies that our species has always been capable of achieving great feats when given access to the right knowledge and resources. This revelation could inspire us to push the boundaries of our own capabilities and strive for even greater accomplishments in the future.

One of the most intriguing aspects of the Ancient Alien Theory is the potential for uncovering previously unrecognized connections between ancient cultures. The similarities in the technologies and techniques used in various regions of the world suggest that there may have been a common source of knowledge, potentially provided by extraterrestrial visitors. By examining these connections, we can gain a more comprehensive understanding of the global exchange of ideas and the shared experiences of ancient civilizations.

Finally, by entertaining the possibility of ancient alien interactions, we are encouraged to broaden our perspective on our place in the cosmos. If extraterrestrial beings have indeed played a role in our past, it suggests that we are not alone in the universe and that our history is intertwined with a vast network of intelligent life. This realization can help us to redefine our sense of identity and purpose, inspiring us to seek out new opportunities for exploration and discovery in the vast expanse of space.

In conclusion, the Ancient Alien Theory offers a captivating alternative view of human history, one that challenges conventional explanations and opens up new avenues for inquiry and understanding. The evidence presented in this chapter, while not definitive, provides a compelling case for considering the potential influence of extraterrestrial beings in our past. By embracing this perspective, we can unlock new insights into the origins of human civilization and our place in the cosmos, ultimately enriching our understanding of the world and the universe beyond.

Chapter 14

Lost Civilizations and Adventures

Journeying Beyond Our Realm

The ancient civilization of Atlantis, as described by Plato, showcased a level of knowledge and architectural prowess that surpassed its contemporaries. The city was said to be composed of concentric rings of water and land, with a central island containing a palace and a massive temple dedicated to Poseidon. Advanced hydraulic systems, including aqueducts and canals, ensured the efficient distribution of water throughout the city. Proponents of the Ancient Alien Theory suggest that such sophisticated urban planning and construction techniques could have been the result of knowledge and technology shared by extraterrestrial visitors.

Atlantis was believed to possess advanced technology that was far ahead of its time. According to the descriptions provided by Plato and other ancient sources, Atlantean society had access to a powerful energy source called "orichalcum," which illuminated the city and powered various machines. Additionally, the Atlanteans were said to possess advanced transportation systems, including flying machines known as "Vailixi." The Ancient Alien Theory proposes that these technological marvels could have been introduced to the Atlanteans by extraterrestrial beings, who either provided the technology directly or guided the Atlanteans in its development.

The advanced culture and society of Atlantis, as depicted in ancient texts, has led some researchers to suggest that extraterrestrial influence may have played a role in shaping the civilization. The Atlanteans were described as having a deep understanding of astronomy, mathematics, and other sciences, which may have been taught to them by ancient aliens. Additionally, their social structure was said to be based on wisdom, cooperation, and the pursuit of higher knowledge, which could indicate the influence of more advanced beings guiding their development.

The sudden disappearance of Atlantis, which was said to have sunk beneath the waves in a single day and night, has been a source of fascination and speculation for centuries. Some proponents of the Ancient Alien Theory suggest that the cataclysmic event that led to the destruction of Atlantis could have been caused by the misuse of advanced alien technology or even by direct extraterrestrial intervention. They argue that ancient aliens may have decided to remove Atlantis from the world stage, either due to the civilization's growing power or as a consequence of the Atlanteans straying from the path set by their extraterrestrial mentors.

The legend of Atlantis, with its advanced knowledge, technology, and culture, serves as a compelling example of how the Ancient Alien Theory can provide alternative explanations for the mysteries of human history. By considering the possibility that extraterrestrial beings played a role in the development and ultimate destruction of Atlantis, researchers can explore new avenues of inquiry and challenge conventional understanding. The enduring fascination with Atlantis and its potential connections to ancient aliens serves as a testament to the impact of these ideas on our collective imagination and our quest for answers about our past.

The Bimini Road, located in the Bahamas, is a peculiar underwater formation consisting of large, rectangular limestone blocks. Some researchers argue that this structure could be part of the lost city of

Atlantis, as its arrangement suggests a potentially man-made origin. Proponents of the Ancient Alien Theory suggest that the advanced engineering skills required to create such a formation could have been imparted by extraterrestrial visitors.

Another underwater discovery that has been linked to Atlantis is the submerged city found off the coast of Cuba. In 2001, a team of researchers discovered mysterious structures deep beneath the ocean's surface, including what appear to be pyramids, roads, and buildings. Although the exact origin of these structures remains unknown, some Ancient Alien Theory proponents believe that they could be remnants of Atlantis, further suggesting a connection between this lost civilization and extraterrestrial beings.

The Emerald Tablets of Thoth, an ancient text believed to be written by the Egyptian god Thoth (also known as Hermes Trismegistus in Greek mythology), is said to contain a wealth of esoteric knowledge and wisdom. Proponents of the Ancient Alien Theory argue that the advanced knowledge found within the Emerald Tablets could be evidence of extraterrestrial contact, as the text describes advanced technologies, such as the construction of pyramids, that seem beyond the capabilities of ancient civilizations.

Similarly, the Oera Linda Book, a controversial manuscript written in Old Frisian, is believed by some to be a chronicle of the ancient Frisian people and their interactions with a lost civilization known as Atland, which has been associated with Atlantis. This text, while disputed by many scholars, is thought by some Ancient Alien Theory proponents to contain evidence of extraterrestrial involvement in the development of these ancient cultures.

The similarities between the architectural styles and myths of various ancient cultures around the world have been cited as evidence for the existence of Atlantis and its possible connections with extraterrestrial beings. For example, the architectural similarities between the pyramids found in Egypt, Mexico, and Central America suggest a shared knowledge and technology that could be attributed to ancient alien contact.

Moreover, many ancient cultures have myths and legends describing a great flood or catastrophe that destroyed a highly advanced civilization. These shared stories, found in cultures as diverse as the Sumerians, Greeks, and Native Americans, could be interpreted as evidence of a common ancient history, potentially involving contact with otherworldly beings.

By examining the anomalous discoveries, ancient texts, and shared cultural connections, proponents of the Ancient Alien Theory argue that the lost civilization of Atlantis may have had a close relationship with extraterrestrial beings. This hypothesis suggests that the advanced knowledge and technology possessed by the Atlanteans could have been imparted by ancient astronauts, who may have played a crucial role in the development and eventual disappearance of this enigmatic civilization.

The concept of Lemuria as a lost continent can be traced back to the 19th-century scientific debates surrounding the geological and biological history of the Indian Ocean region. At the time, scientists were puzzled by the presence of similar species of plants and animals on islands separated by vast distances. To account for this, some proposed the existence of a now-submerged landmass, which they named Lemuria, that had once connected these disparate regions. Although this idea has since been largely discredited in the scientific community, it has continued to capture the imagination of many who are interested in the possibility of an advanced, ancient civilization that once thrived in this region.

The concept of Lemuria as a highly advanced civilization emerged in the late 19th and early 20th centuries, primarily within the realms of esotericism and alternative history. Various authors and researchers have presented theories and speculations about the nature of this lost continent, suggesting that it was home to a technologically and spiritually advanced society. Proponents of the Ancient Alien Theory argue that the high level of advancement demonstrated by the Lemurians could be the result of contact with extraterrestrial beings, who shared their knowledge and technology with this ancient civilization.

Supporters of the Ancient Alien Theory point to various architectural and archaeological discoveries as evidence for the existence of Lemuria and its advanced civilization. For instance, the enigmatic ruins of Nan Madol, an ancient city in Micronesia, are often cited as a potential remnant of

Lemurian architecture. This complex of megalithic structures, built upon a series of artificial islets, demonstrates an impressive level of engineering skill and an architectural style that has yet to be fully understood. Similarly, the underwater Yonaguni Monument near Japan, a massive rock formation with seemingly artificial features, has been proposed as a possible remnant of the lost continent of Lemuria.

In addition to the archaeological evidence, proponents of the Ancient Alien Theory suggest that various cultural and mythological traditions may contain clues about the existence of Lemuria and its extraterrestrial connections. For example, the Tamil people of southern India have a rich literary tradition that includes references to a lost continent known as Kumari Kandam, which some researchers have linked to Lemuria. Additionally, the concept of the "Seven Rishis" or ancient sages in Hindu mythology has been interpreted by some as evidence of advanced knowledge imparted to the Lemurians by extraterrestrial visitors.

The idea of Lemuria as a lost, advanced civilization has had a lasting impact on popular culture and alternative historical research, particularly within the context of the Ancient Alien Theory. By exploring the potential connections between this enigmatic continent and extraterrestrial beings, proponents of the theory challenge conventional explanations and invite us to consider the possibility that our understanding of human history may be far more complex and interconnected than we have previously imagined. As researchers continue to investigate the mysteries surrounding Lemuria and its potential links to ancient aliens, this lost continent will undoubtedly remain a captivating and thought-provoking topic of inquiry.

Lemurian creation myths, which are prevalent in various esoteric teachings and literature, often allude to the idea that Lemuria was established by extraterrestrial beings. According to these accounts, ancient aliens arrived on Earth and created the Lemurian civilization by either genetically engineering the existing human population or by introducing their own genetic material to form a hybrid species. This notion of an extraterrestrial genesis for Lemuria aligns with the broader Ancient Alien Theory, as it suggests that these advanced beings played a pivotal role in shaping the development of human civilization.

The Lemurian civilization is often described as possessing advanced knowledge and technology that surpassed the capabilities of other ancient societies. Proponents of the Ancient Alien Theory suggest that this advanced knowledge could be attributed to contact with extraterrestrial beings who shared their expertise in various fields, such as agriculture, medicine, astronomy, and engineering. This exchange of information would have enabled the Lemurians to develop a sophisticated culture that left a lasting impact on human history.

One of the most compelling arguments for the connection between Lemuria and ancient aliens lies in the architectural achievements and megalithic structures attributed to this lost civilization. Sites like Nan Madol in Micronesia and the underwater Yonaguni Monument near Japan exhibit remarkable construction techniques and precise engineering that seem to defy the capabilities of ancient civilizations. Ancient Alien Theory proponents argue that these feats were made possible through the knowledge and technology shared by extraterrestrial visitors, who may have provided guidance or even direct assistance in the construction of these impressive structures.

Numerous legends and accounts of Lemuria and its inhabitants can be found in various cultures and spiritual traditions around the world. Some of these stories describe the Lemurians as a race of highly advanced beings, possessing unique abilities and knowledge that set them apart from other ancient civilizations. In addition to these oral and written accounts, various artifacts and relics purportedly linked to Lemuria have been discovered over the years. These items, which include statues, carvings, and inscriptions, are often cited by Ancient Alien Theory proponents as evidence of the advanced nature of Lemurian society and its connection to extraterrestrial beings.

The idea of Lemuria and its extraterrestrial connections has had a profound impact on modern esoteric thought and alternative history research. Many contemporary spiritual and philosophical movements, such as Theosophy and New Age teachings, incorporate the concept of Lemuria and its ancient alien origins into their belief systems. These modern interpretations of Lemurian history and culture serve to further promote the Ancient Alien Theory, as they continue to explore the potential influence of extraterrestrial beings on the development of human civilization throughout the ages.

Many ancient cultures have creation myths that feature cosmic deities or beings from the sky who played a significant role in the formation of the world and the origins of human civilization. For instance, the Enuma Elish, a Babylonian creation myth, tells the story of the god Marduk, who defeated the primordial chaos monster Tiamat and created the world from her remains. Similarly, the Egyptian myth of the Ennead describes the creation of the world by the god Atum, who emerged from the primordial waters of Nun. Proponents of the Ancient Alien Theory argue that these cosmic deities could represent extraterrestrial beings who played a role in the development of human civilization.

In many ancient myths, gods or celestial beings are described as teachers and advisors, bestowing knowledge and wisdom upon human beings. For example, the Greek myth of Prometheus tells the story of a Titan who defied the gods by stealing fire from the heavens and giving it to humanity, symbolizing the sharing of knowledge and technology. In Sumerian mythology, the Anunnaki are said to have taught humans the arts of civilization, including agriculture, writing, and architecture. Ancient Alien Theory proponents suggest that these myths may be based on encounters with extraterrestrial beings who shared their advanced knowledge with early human societies.

Stories of a great flood that wiped out previous civilizations are found in numerous ancient cultures, including the Sumerians, Egyptians, Greeks, and Native Americans. The Epic of Gilgamesh, one of the oldest known literary works, tells the story of a great flood sent by the gods to destroy humanity, with one man, Utnapishtim, being instructed to build an ark to save a remnant of life. Proponents of the Ancient Alien Theory argue that these flood stories may represent a shared memory of an actual global cataclysm, potentially triggered or influenced by extraterrestrial intervention.

Many ancient myths and legends feature beings from the sky or celestial vehicles, which could be interpreted as evidence of extraterrestrial contact. In Hindu texts, the Vimanas are described as flying machines piloted by gods, capable of incredible feats of speed and maneuverability. Similarly, the Book of Enoch, an apocryphal Jewish text, tells of the Watchers, angelic beings who descended from the heavens to teach and guide humanity. Proponents of the Ancient Alien Theory argue that these stories may be based on encounters with ancient astronauts and their advanced technology.

The similarities between the myths and legends of various cultures serve as a compelling argument for the Ancient Alien Theory, suggesting that early human societies may have had shared encounters with extraterrestrial beings. These myths not only demonstrate a commonality in the themes of creation, destruction, and divine intervention but also provide intriguing hints of advanced knowledge and technology that could have been bestowed upon humanity by ancient astronauts. As researchers continue to explore the connections between these myths and the possibility of extraterrestrial contact, the shared history of lost civilizations and ancient aliens will remain a fascinating area of study.

The Great Pyramid of Giza, one of the most iconic symbols of ancient Egypt, has long been a subject of fascination for researchers and Ancient Alien Theory proponents alike. The pyramid's precise alignment with the cardinal directions, its intricate internal design, and the massive size of the stones used in its construction have led some to believe that the Egyptians may have received assistance from extraterrestrial beings. The level of engineering and mathematical knowledge required to build such a precise and massive structure seems to exceed the capabilities of the ancient Egyptians, as no written records or blueprints have ever been found to explain their methods.

Pre-Columbian cultures in South America, such as the Incas, are known for their advanced metallurgical skills. They were capable of producing complex metal objects, including gold and silver alloys, using techniques that seem to be far ahead of their time. For example, the ancient Mesoamerican civilization of the Toltecs was known for its remarkable metalwork, producing intricate gold artifacts that demonstrate a high level of technical skill. Proponents of the Ancient Alien Theory argue that such advanced metallurgy could be the result of extraterrestrial guidance or intervention, as it is difficult to explain how these cultures developed such sophisticated techniques independently.

The Mayan civilization, which thrived in Central America between 2000 BCE and 1500 CE, was renowned for its sophisticated understanding of astronomy. The Mayans developed an advanced calendar system that accurately tracked celestial movements, including the orbits of planets and the cycles of the Moon and Venus. They also built monumental structures, such as the El Castillo pyramid at Chichen Itza, which were designed to align with astronomical events like the equinoxes and solstices. The accuracy and complexity of the Mayan astronomical knowledge have led some Ancient Alien Theory supporters to propose that this civilization may have been in contact with extraterrestrial beings who shared their advanced understanding of the cosmos.

Megalithic structures, such as Stonehenge in England, the Moai statues of Easter Island, and the massive stones at Baalbek in Lebanon, have long puzzled researchers and fueled speculation about the existence of ancient aliens. The sheer size and weight of the stones used in these constructions, combined with the precision and complexity of their arrangements, have led many to question how such feats could have been accomplished without the help of advanced technology. Proponents of the Ancient Alien Theory argue that the construction of these megalithic structures could be evidence of extraterrestrial involvement in human history, as the techniques used seem to defy the capabilities of the civilizations responsible for their creation.

The advanced technology and knowledge displayed by lost civilizations, as evidenced by their archaeological remains, have led many to consider the possibility of extraterrestrial intervention. Ancient Alien Theory proponents argue that the achievements of these civilizations, such as the precise construction of megalithic structures, advanced metallurgy, and sophisticated astronomical knowledge, were beyond their capabilities and can only be explained by the involvement of otherworldly beings. By exploring the potential connections between ancient aliens and the advanced technology of lost civilizations, researchers in the field challenge conventional explanations and invite us to reevaluate our understanding of human history and the broader implications of extraterrestrial life.

The Anunnaki, central figures in Sumerian mythology, are often cited as evidence of ancient alien influence on spiritual practices and beliefs. Described as a pantheon of gods that descended from the heavens, the

Anunnaki were said to have played a pivotal role in shaping human civilization. They were credited with providing advanced knowledge in agriculture, law, and even genetic manipulation. Proponents of the Ancient Alien Theory argue that the Anunnaki's advanced abilities and skyward origin point to the possibility of them being extraterrestrial visitors who were revered as gods by the Sumerians.

The Vimanas, described in ancient Hindu texts such as the Mahabharata and the Ramayana, are another example of potential ancient alien influence on spiritual beliefs. Vimanas were described as flying machines or celestial chariots piloted by the gods, capable of traveling vast distances and even venturing into outer space. They were said to be equipped with advanced weaponry and technology, far beyond the capabilities of ancient human civilization. Ancient Alien Theory proponents argue that the detailed descriptions of Vimanas and their advanced technology suggest that these flying machines could have been extraterrestrial spacecraft operated by ancient astronauts.

Quetzalcoatl, a prominent deity in Mesoamerican cultures such as the Aztecs and the Maya, is another example of a spiritual figure that could be linked to ancient alien contact. Known as the Feathered Serpent, Quetzalcoatl was associated with wisdom, knowledge, and advanced technology. He was said to have brought the gift of agriculture and other vital knowledge to the people of Mesoamerica. Ancient Alien Theory supporters propose that Quetzalcoatl may have been an ancient astronaut who visited Earth and shared advanced knowledge with the Mesoamerican people, who then revered him as a god.

The Dogon people of Mali, West Africa, have long been of interest to Ancient Alien Theory researchers due to their complex astronomical knowledge, particularly regarding the Sirius star system. The Dogon possessed accurate knowledge about the existence of Sirius B, a white dwarf star that is invisible to the naked eye, long before its discovery by modern astronomers. They attribute this knowledge to their ancestors' contact with extraterrestrial beings called the Nommo, who are said to have visited Earth from the Sirius star system. This example demonstrates how ancient alien contact may have influenced the spiritual beliefs and practices of the Dogon tribe.

The phenomenon of cargo cults in the Pacific Islands provides a modern-day example of how advanced technology can be perceived as divine by less technologically advanced cultures. During World War II, indigenous islanders observed the arrival of planes and military cargo, which they believed to be gifts from the gods. As a result, they developed spiritual practices and rituals centered around the belief that these gods would return with more gifts. Ancient Alien Theory proponents suggest that this phenomenon mirrors the way ancient civilizations may have perceived advanced extraterrestrial visitors and their technology as divine, incorporating them into their spiritual beliefs and practices.

The advent of satellite imagery has revolutionized the search for lost civilizations and their potential connections to ancient aliens. High-resolution images captured by satellites have allowed researchers to identify previously undiscovered archaeological sites, even in the most remote and inaccessible regions of the world. These images have led to the discovery of ancient settlements, temples, and other structures that may provide clues about the interactions between ancient cultures and extraterrestrial beings. Some researchers argue that certain geometric patterns and alignments observed in these sites suggest advanced knowledge and technology that could have originated from ancient alien visitors.

Underwater exploration has opened up new avenues for the investigation of lost civilizations and their possible connections to ancient aliens. Advanced submersible vehicles and diving equipment have enabled researchers to explore submerged structures and artifacts that were once beyond our reach. These discoveries have included underwater ruins, such as the Yonaguni Monument near Japan and the submerged city off the coast of Cuba, which some researchers believe may be remnants of lost civilizations with extraterrestrial connections. The advanced technology and design evident in these underwater structures have led proponents of the Ancient Alien Theory to argue that they could be evidence of ancient contact with otherworldly beings.

Remote sensing technologies, such as LiDAR (Light Detection and Ranging), have provided archaeologists with powerful tools for the discovery and analysis of lost civilizations and their potential extraterrestrial connections. By using lasers to create detailed, three-dimensional maps of the Earth's surface, LiDAR has enabled researchers to uncover hidden archaeological sites and gain new insights into the advanced technology and knowledge of ancient cultures. In some cases, the data gathered through remote sensing has revealed the presence of previously unknown structures and settlements, which may offer further evidence of ancient alien contact.

The search for lost civilizations and their extraterrestrial connections has increasingly become a multidisciplinary endeavor, involving experts from various fields such as archaeology, anthropology, geology, and even astronomy. By combining their expertise and knowledge, these researchers have been able to piece together a more comprehensive picture of the ancient world and its potential interactions with extraterrestrial beings. This collaborative approach has led to new theories and interpretations of archaeological evidence, as well as innovative methods for analyzing and dating ancient artifacts and structures.

As technology continues to advance and our understanding of the ancient world expands, the search for lost civilizations and their potential connections to ancient aliens is likely to yield even more groundbreaking discoveries. By employing cutting-edge tools and techniques, researchers will be better equipped to unravel the mysteries of the past and uncover evidence that may support the Ancient Alien Theory. As we continue to explore the depths of our history and the possibilities of extraterrestrial life, the quest for lost civilizations and their otherworldly connections will undoubtedly remain a fascinating and thought-provoking journey for generations to come.

The Ancient Alien Theory proposes that some natural disasters that led to the decline or destruction of ancient civilizations could have been caused or exacerbated by extraterrestrial beings using advanced technology. For instance, some researchers suggest that the Great Flood, a catastrophic event mentioned in various ancient texts, may have been the result of advanced weather manipulation technology utilized by ancient aliens. This could have been an attempt to control or reshape human populations, or even as an unintended consequence of their interference with Earth's natural processes.

Another possibility considered by proponents of the Ancient Alien Theory is that conflicts between different extraterrestrial factions may have had a significant impact on the fate of lost civilizations. These conflicts could have manifested as wars fought using advanced weaponry, resulting in widespread destruction and the collapse of these ancient societies. Alternatively, political or ideological disagreements between extraterrestrial factions could have led to a withdrawal of their support and technology, leaving these civilizations vulnerable to decline.

Ancient Alien Theory researchers also speculate that human allies of various extraterrestrial factions might have played a role in the decline of lost civilizations. These human allies, potentially members of the ruling elite or priestly class, could have been entrusted with advanced knowledge and technology provided by their extraterrestrial benefactors. However, as conflicts or power struggles emerged, these human allies may have turned against one another, causing societal disintegration and the eventual collapse of their civilizations.

Another area of interest within the Ancient Alien Theory is the potential role of genetic manipulation in the decline of lost civilizations. Proponents of the theory argue that ancient aliens may have conducted genetic experiments on human populations or even created hybrid species. The unintended consequences of such manipulation, such as the emergence of genetic disorders or the destabilization of ecosystems, could have contributed to the decline or collapse of these ancient societies.

By examining the potential role of ancient aliens in the decline of lost civilizations, researchers in the field aim to better understand the complex interactions between extraterrestrial beings and human societies throughout history. These investigations provide valuable insights into the potential risks and consequences of advanced technology, as well as the importance of responsible stewardship of such knowledge. Ultimately, the study of the decline of lost civilizations and their potential connections to ancient aliens serves as a reminder of the need for caution and wisdom as humanity continues to explore the cosmos and unlock the secrets of advanced technology.

The Ancient Alien Theory has forced historians, archaeologists, and anthropologists to reconsider traditional explanations for the achievements of lost civilizations. By proposing that extraterrestrial beings played a significant role in shaping these cultures, the theory introduces an alternative perspective on human history, one that goes beyond the scope of conventional historical narratives. This approach has led to a reevaluation of archaeological evidence and ancient texts, prompting new interpretations and theories about the origins and development of these enigmatic civilizations.

The possibility that ancient aliens influenced the development of lost civilizations has inspired researchers to explore cutting-edge scientific theories and technologies. By examining the potential evidence for extraterrestrial contact in our past, scientists have been prompted to investigate fields such as astrobiology, advanced propulsion systems, and genetic engineering. In this way, the Ancient Alien Theory has not only expanded our understanding of the past but has also driven innovation and progress in various scientific disciplines.

The concept of ancient aliens and their connections to lost civilizations has encouraged people to view human history through a more inclusive and interconnected lens. By highlighting the similarities between different cultures and their shared experiences with otherworldly beings, the Ancient Alien Theory fosters a greater appreciation for the diverse traditions, beliefs, and achievements of societies around the world. This more holistic perspective on our shared human past promotes cultural exchange, empathy, and a deeper understanding of our global heritage.

The idea of ancient aliens and their role in the development of lost civilizations has captured the imagination of millions, influencing literature, film, television, and other forms of popular culture. Works such as Erich von Däniken's "Chariots of the Gods?" and the History Channel's "Ancient Aliens" series have popularized the theory and brought it to a wider audience. This widespread fascination with the idea of ancient extraterrestrial contact has stimulated creative thinking and inspired countless stories and artistic expressions that explore the potential implications of such encounters.

As new evidence and discoveries continue to emerge, the Ancient Alien Theory will likely play a significant role in shaping future research and exploration in the fields of archaeology, history, and anthropology. The ongoing quest to uncover the mysteries of lost civilizations and their potential connections to extraterrestrial beings will undoubtedly drive further investigation into unexplained archaeological findings, ancient texts, and cultural practices. By challenging conventional explanations and pushing the boundaries of our understanding of the past, the legacy of lost civilizations and the Ancient Alien Theory will continue to inspire curiosity, debate, and discovery for generations to come.

Chapter 15

Stories and Legends about Ancient Aliens

The Stories about Ancient Alien Environments and Species

Throughout human history, countless myths, legends, and stories have described encounters with otherworldly beings and their extraordinary environments. These narratives often contain elements that suggest the existence of advanced alien species, which may have influenced the development of human civilizations. By examining these stories and legends in detail, we can gain valuable insights into the potential existence of ancient alien environments and species.

The Sumerians believed that the Anunnaki played a significant role in the formation and development of their society. These beings were said to have taught the Sumerians various aspects of civilization, including agriculture, architecture, and the arts. Ancient astronaut theorists argue that this knowledge transfer is evidence of extraterrestrial intervention, as the rapid development of the Sumerian civilization is difficult to explain through purely terrestrial means.

Zecharia Sitchin, a prominent ancient astronaut theorist, has extensively researched the Sumerian accounts of the Anunnaki and their home planet, Nibiru. According to Sitchin, Nibiru is the twelfth planet in our solar system, with an elongated orbit that brings it close to Earth every 3,600 years. He suggests that the Anunnaki's visits to Earth coincided with Nibiru's periodic approach, providing opportunities for the exchange of knowledge and technology between the two civilizations.

Some ancient astronaut theorists propose that the Anunnaki were not only responsible for imparting knowledge and technology to the Sumerians but also played a role in the creation of humans. According to this theory, the Anunnaki genetically engineered humans by combining their own DNA with that of early hominids. This process, known as genetic manipulation or hybridization, is said to have resulted in the creation of Homo sapiens as a species capable of advanced thought and communication.

The influence of the Anunnaki can also be seen in the art and architecture of the Sumerian civilization. Many Sumerian artifacts and structures feature depictions of winged beings and other symbols commonly associated with the Anunnaki. Ancient astronaut theorists argue that these artistic representations are evidence of the Sumerians' contact with extraterrestrial beings and their attempts to document and honor their otherworldly benefactors.

The stories of the Anunnaki continue to captivate and inspire researchers and enthusiasts of the Ancient Alien Theory. The possibility that these beings were, in fact, extraterrestrial visitors who played a crucial role in the development of human civilization challenges conventional historical narratives and expands our understanding of humanity's place in the cosmos. By studying the Sumerian accounts of the Anunnaki and their interactions with early human societies, we can gain valuable insights into the potential existence of advanced extraterrestrial life and the profound impact they may have had on our ancestors.

The Mahabharata, an ancient Indian epic, contains numerous accounts of Vimanas used by gods and heroes during battles and other significant events. These flying machines were said to possess incredible maneuverability and speed, with some being described as having the ability to become invisible or change their shape. Ancient alien theorists suggest that the extraordinary capabilities of the Vimanas, as depicted in the Mahabharata, point to the existence of advanced extraterrestrial technology.

In the Ramayana, another ancient Indian epic, the hero Rama is said to have utilized a flying machine called the Pushpaka Vimana. This particular Vimana was described as a self-propelled vehicle that could change its shape and size according to the needs of its passengers. The Pushpaka Vimana was also said to be able to travel vast distances, even between different realms, which ancient alien theorists argue is suggestive of interstellar or interdimensional travel capabilities.

Some ancient alien theorists propose that the pilots of the Vimanas described in ancient Indian texts were not just gods or mythological figures, but actual ancient astronauts – human or extraterrestrial – who had access to advanced technology. These ancient astronauts may have been responsible for spreading their knowledge and technology to the people of ancient India, contributing to the development of the advanced civilization that thrived during that time.

The Vaimanika Shastra is an early 20th-century Sanskrit text that claims to be a transcription of an ancient treatise on the construction and operation of Vimanas. The text provides detailed descriptions of various types of Vimanas and their propulsion systems, as well as advanced materials and weaponry. While the authenticity of the Vaimanika Shastra is debated, ancient alien theorists argue that the text may contain remnants of ancient knowledge regarding extraterrestrial technology that were passed down through generations.

Ancient alien proponents often draw parallels between the Vimanas described in ancient Indian texts and modern-day UFO sightings. They argue that the striking similarities between these accounts, such as the description of advanced propulsion systems, maneuverability, and the ability to change shape, support the idea that extraterrestrial beings have been visiting Earth throughout human history. The Vimanas, according to this theory, may represent early encounters with these otherworldly visitors and their advanced technology.

The presence of Vimanas in ancient Indian texts has fueled speculation about the existence of advanced extraterrestrial technology and its influence on ancient human civilizations. While definitive proof remains elusive, the detailed descriptions of Vimanas and their capabilities provide a fascinating insight into the potential encounters between our ancestors and advanced alien visitors. As we continue to explore the mysteries of our past, the Vimanas of ancient India serve as a compelling reminder of the possibilities that lie within the vast expanse of the cosmos.

In Greek mythology, the Hyperboreans inhabited a remote and idyllic region far to the north, beyond the reach of the cold north wind Boreas. This land was described as a paradise, where the sun shone constantly, and the inhabitants enjoyed eternal youth and happiness. The location of this mythical land remains a mystery, with some ancient sources placing it near the Arctic Circle, while others suggest it could be located in the British Isles or the Baltic region.

The Hyperboreans were believed to possess advanced knowledge and technology that allowed them to live incredibly long lives and maintain their utopian society. Their mastery over various arts and sciences was said to be unparalleled, and they were known for their extraordinary skills in music, medicine, and other disciplines. Ancient astronaut proponents argue that this advanced knowledge could have been the result of contact with extraterrestrial visitors, who shared their advanced technology and knowledge with the Hyperboreans.

The mythology of the Hyperboreans is closely tied to the Greek gods, particularly Apollo, who was said to visit the Hyperborean homeland every winter. According to legend, the Hyperboreans enjoyed a close relationship with the gods, regularly exchanging gifts and participating in sacred rituals. This connection between the Hyperboreans and the divine further supports the ancient astronaut theory, as it suggests that these advanced beings may have been perceived as gods or divine messengers by the ancient Greeks.

Ancient astronaut theorists contend that the Hyperboreans may have had a profound influence on the development of ancient Greek civilization, sharing their advanced knowledge and technology with the people of Greece. This is supported by the fact that the Hyperboreans were often associated with the Muses, the goddesses of inspiration and creativity, who were believed to have imparted their wisdom to the great poets, philosophers, and scientists of ancient Greece.

Given the descriptions of the Hyperborean homeland, their advanced knowledge and technology, and their close relationship with the gods, ancient astronaut proponents argue that the Hyperboreans may have been extraterrestrial beings who visited Earth in ancient times. The idyllic, otherworldly environment they inhabited could be indicative of their advanced technology and ability to create a utopian society, which was beyond the reach of the contemporary human civilizations.

The Hyperboreans in Greek mythology represent a fascinating case for the ancient astronaut theory. Their advanced knowledge, technology, and the utopian society they inhabited suggest the possibility of extraterrestrial origins. By examining these ancient myths and legends, we can explore the potential existence of ancient alien visitors and their influence on human civilization throughout history.

The Dogon tribe's detailed knowledge of the Sirius star system is remarkable considering their lack of access to modern astronomical equipment. Their accurate understanding of Sirius B, a white dwarf star that is invisible to the naked eye, has led ancient astronaut theorists to conclude that this information was provided to them by extraterrestrial beings who visited the Dogon people. Furthermore, the Dogon also possessed knowledge of the 50-year orbital period of Sirius B around its larger companion star, Sirius A, which is another impressive detail given the limitations of their observational capabilities.

The Dogon's stories of the Nommo, the amphibious beings from the Sirius system, include accounts of advanced knowledge that was imparted to the tribe in areas such as agriculture and mathematics. This knowledge enabled the Dogon people to develop sophisticated farming practices, helping them to thrive in the harsh environment of West Africa. The Nommo's teachings also covered complex mathematical concepts, which allowed the Dogon to understand and predict various celestial phenomena. Ancient astronaut theorists argue that this advanced knowledge, seemingly beyond the reach of the Dogon people at the time, is a strong indicator of extraterrestrial influence.

The Dogon's oral tradition describes the arrival of the Nommo in a celestial ark, which descended from the heavens and landed on Earth. These accounts include vivid descriptions of the Nommo's appearance, which combines humanoid and amphibious characteristics. According to the Dogon, the Nommo taught them vital skills and provided them with invaluable knowledge before eventually returning to the Sirius star system. Ancient astronaut theorists propose that these stories are an accurate record of an encounter between the Dogon people and an advanced extraterrestrial species.

The Dogon's culture includes various rituals and artifacts that reflect their beliefs in the Nommo and their connection to the Sirius star system. For example, the Dogon's ceremonial masks, which are used during their traditional dances, are believed to represent the Nommo's appearance. Additionally, their sacred architecture, such as the toguna, a communal meeting place, incorporates complex geometric patterns that are said to symbolize the celestial knowledge passed down by the Nommo. Ancient astronaut theorists argue that these cultural elements provide further evidence of the Dogon's contact with extraterrestrial visitors.

The Dogon tribe's stories of the Nommo and their advanced knowledge of the Sirius star system offer a compelling case for the existence of ancient aliens. These accounts suggest that the Dogon people had direct contact with extraterrestrial beings who shared their advanced knowledge with the tribe, enabling them to thrive and develop a rich cultural tradition. This connection between the Dogon and the Nommo serves as a powerful example of how ancient alien interactions may have shaped human history and our understanding of the universe.

Among the Lakota Sioux, a prominent Native American tribe, there are stories of the Star People, who are said to have descended from the heavens to visit the tribe in ancient times. These celestial beings are believed to have shared wisdom about sacred ceremonies, healing practices, and even the construction of the iconic tipi, which closely resembles some modern interpretations of UFOs. The Lakota Sioux's reverence for the Star

People is evident in their spiritual practices and oral traditions, which further strengthens the ancient alien theory's connection to this tribe.

The Iroquois people have a creation story that tells of a Sky Woman who descended from the heavens and gave birth to twin sons on Earth. This legend is significant because it suggests that the Iroquois people believe their origins are connected to celestial beings. Ancient astronaut theorists interpret the Sky Woman as an extraterrestrial visitor who played a crucial role in the creation of the Iroquois people, providing them with guidance and knowledge from the stars.

The Cherokee Nation has a rich cultural history that includes legends of beings from the Pleiades star cluster. According to these stories, the Pleiadians visited the Cherokee people in ancient times, providing them with knowledge of agriculture, language, and other essential skills. The Cherokee's close association with the Pleiades is evident in their spiritual beliefs and traditional ceremonies. Ancient astronaut theorists see these connections as further evidence of extraterrestrial contact with Native American tribes.

Similar to the Hopi people, the Zuni tribe of the southwestern United States also has a deep-rooted belief in Kachina spirits, who are thought to have come from the sky. These otherworldly beings are believed to have provided the Zuni people with advanced knowledge and wisdom, which has been passed down through generations. Ancient astronaut theorists propose that the Kachina spirits were extraterrestrial visitors who interacted with the Zuni tribe, influencing their cultural and spiritual development.

Native American legends of the Star People provide intriguing evidence for ancient alien contact with indigenous tribes. These stories, which span various tribes and geographical locations, consistently depict celestial beings who descended from the sky to share advanced knowledge with the Native Americans. The presence of these legends across diverse Native American cultures lends credibility to the ancient astronaut theory and suggests that the Star People played a crucial role in shaping the beliefs and practices of these ancient civilizations.

The Hopi people believed that Kachinas were divine messengers who played a crucial role in their spiritual and cultural practices. Acting as intermediaries between the human and celestial realms, Kachinas not only imparted advanced knowledge but also represented the forces of nature, which the Hopi revered and sought to understand. The notion that these beings descended from the sky suggests an extraterrestrial origin, further supported by their otherworldly attributes and capabilities.

An essential aspect of Hopi culture is the performance of Kachina dances and rituals. During these ceremonies, participants wear masks and costumes that represent the various Kachinas, aiming to channel the spirits of these otherworldly beings. The elaborate nature of the masks and costumes, often adorned with feathers and intricate patterns, may be seen as an attempt to emulate the appearance of advanced alien visitors. The Kachina dances serve as a form of communication with the celestial beings, with participants seeking their guidance and wisdom.

The Hopi people are renowned for their sophisticated agricultural practices, which have enabled them to thrive in the arid climate of the southwestern United States. Ancient astronaut theorists suggest that the advanced knowledge and techniques employed by the Hopi were gifted to them by the Kachinas, who possessed a deep understanding of Earth's ecosystems and natural processes. The Kachinas may have shared their expertise in water conservation, crop rotation, and soil management, enabling the Hopi to develop sustainable agricultural systems that have endured for centuries.

The Hopi people possess a deep understanding of astronomy, which they have used to track celestial events and develop a calendar system. Their knowledge of the night sky has been attributed to the teachings of the Kachinas, who are believed to have shared their celestial wisdom with the Hopi. Ancient astronaut theorists argue that the Kachinas' knowledge of the cosmos is indicative of their extraterrestrial origins, as they would have possessed a unique perspective on the universe from their celestial vantage point.

Ancient astronaut proponents maintain that the Kachinas' influence on Hopi culture extends beyond the realms of agriculture and astronomy. They argue that the Kachinas were instrumental in shaping the social, political, and artistic aspects of Hopi society, imparting their advanced knowledge and technologies to enrich the lives of the people they encountered. The Kachinas' impact on Hopi culture serves as a testament to the profound connection between the Hopi people and these otherworldly beings, further supporting the theory of ancient alien contact.

Ancient Chinese literature contains numerous references to the Dragon Kings, often in the context of their interactions with human rulers and sages. One such example is found in the classic text, "The Classic of Mountains and Seas," which describes the Dragon Kings as residing in underwater palaces and ruling over their respective domains. The text also highlights their role in bestowing wisdom and guidance upon human rulers, who would seek their counsel in times of need.

The four Dragon Kings, each associated with one of the cardinal directions, are believed to rule over the four seas surrounding China. These beings, named Ao Guang (East), Ao Qin (South), Ao Run (West), and Ao Shun (North), are said to have the ability to manipulate water, summon rain, and control the weather. Their mastery over the elements is indicative of advanced knowledge and technology, which ancient alien proponents argue is consistent with the capabilities of an extraterrestrial civilization.

The underwater palaces of the Dragon Kings are described as being constructed from precious materials such as jade and gold, with magical properties that enable them to remain hidden from human sight. These palaces are also said to house advanced technologies, such as the "dragon's pearls," which are thought to have the ability to emit light and heal ailments. Ancient astronaut theorists propose that these descriptions are evidence of the advanced technology possessed by the Dragon Kings, possibly pointing to their extraterrestrial origins.

The Dragon Kings are not only depicted as powerful and wise beings in Chinese mythology but are also often portrayed as having a direct influence on Chinese culture. They are believed to have shared their knowledge and technology with humans, assisting in the development of agriculture, art, and architecture. For example, the dragons' ability to control water and summon rain is believed to have played a crucial role in the successful establishment of ancient Chinese agriculture. Proponents of the Ancient Alien Theory argue that this cultural influence is consistent with the actions of an advanced extraterrestrial civilization, seeking to guide humanity's development.

Dragons are deeply ingrained in Chinese culture and are regarded as symbols of power, strength, and good fortune. The reverence for these mythical creatures is often attributed to their association with the Dragon Kings and their otherworldly abilities. Ancient astronaut theorists suggest that the widespread veneration of dragons in Chinese culture is evidence of the impact made by these extraterrestrial beings on ancient Chinese society.

The legends of the Dragon Kings in Chinese mythology provide a compelling basis for the Ancient Alien Theory. The advanced knowledge, technology, and otherworldly abilities attributed to these beings, along with their deep-rooted influence on Chinese culture, suggest the possibility of extraterrestrial contact with ancient Chinese civilizations. While definitive proof remains elusive, the stories of the Dragon Kings offer intriguing insights into the potential existence of ancient alien visitors and their impact on human history.

Quetzalcoatl's depiction as a "Feathered Serpent" is particularly intriguing to ancient astronaut theorists, as it combines the features of a serpent with the plumage of a bird. The serpent, in many cultures, is associated with wisdom, while the bird symbolizes the ability to traverse the sky. This unique combination of features suggests that Quetzalcoatl might represent an extraterrestrial being who possessed both advanced knowledge and the ability to travel through the cosmos.

According to Mesoamerican legends, Quetzalcoatl arrived from the East, bringing with him the knowledge and skills that would shape the development of these ancient civilizations. This arrival is often associated with the appearance of a bright star or comet, which ancient astronaut theorists argue could have been the arrival of an extraterrestrial spacecraft.

Similarly, Quetzalcoatl is said to have departed to the East, promising to return one day. This departure is also associated with celestial phenomena, further supporting the idea that Quetzalcoatl was an alien visitor who traveled to and from Earth using advanced spacecraft.

The teachings attributed to Quetzalcoatl are often seen as evidence of his extraterrestrial origins. He is said to have introduced advanced agricultural techniques, such as the cultivation of maize and the use of the

Mesoamerican calendar, which allowed for precise tracking of time and celestial events. Quetzalcoatl also provided knowledge in mathematics, including the concept of zero, which was not known in other ancient civilizations at the time.

Ancient astronaut theorists argue that the advanced knowledge and technologies introduced by Quetzalcoatl were beyond the capabilities of the indigenous people of Mesoamerica at the time. They believe that these teachings could only have come from an advanced extraterrestrial source, which Quetzalcoatl represented.

The promise of Quetzalcoatl's return is an important aspect of Mesoamerican mythology, and ancient astronaut theorists suggest that this could represent the anticipated return of extraterrestrial visitors. The belief in Quetzalcoatl's return is said to have influenced the Aztec's reaction to the arrival of the Spanish conquistadors, who were initially mistaken for the returning god due to their fair skin and advanced technology.

This connection between Quetzalcoatl's return and the arrival of technologically advanced visitors further reinforces the idea that the legend of Quetzalcoatl may be rooted in ancient encounters with extraterrestrial beings.

The legend of Quetzalcoatl in Mesoamerican mythology provides a compelling basis for the idea that ancient aliens may have visited and influenced the development of these civilizations. The advanced knowledge attributed to Quetzalcoatl, his association with celestial phenomena, and his unique physical appearance all suggest that this mythological figure could represent an extraterrestrial visitor. By examining the Quetzalcoatl legend through the lens of the Ancient Alien Theory, we can gain a deeper understanding of the potential existence of advanced alien species and their interactions with ancient human civilizations.

The various myths, legends, and stories of ancient alien environments and species found across human history offer tantalizing glimpses into the possibility of extraterrestrial contact with our ancestors. These narratives, while not constituting definitive proof of ancient alien visitation, provide a compelling basis for further investigation and speculation.

One notable aspect of these ancient stories is the striking cross-cultural similarities they share, despite being separated by vast geographical distances and cultural barriers. These parallels suggest a common underlying theme or influence, potentially pointing to the existence of ancient alien visitors who interacted with various civilizations across the globe. By examining these stories collectively, we can piece together a more coherent and persuasive narrative that supports the Ancient Alien Theory.

The accounts of advanced knowledge and technology in these ancient myths and legends are particularly intriguing, as they often describe capabilities that would have been unimaginable to the people of the time. These descriptions challenge our conventional understanding of history and suggest that ancient civilizations may have had access to extraterrestrial knowledge or assistance. The Ancient Alien Theory provides a plausible explanation for these advanced technologies, opening up new avenues of research and exploration into our past.

Ancient art and symbolism also play a significant role in supporting the idea of ancient alien contact. Throughout history, various cultures have depicted otherworldly beings and advanced technologies in their artwork, often using similar symbols and motifs. These artistic representations, when viewed in the context of the Ancient Alien Theory, provide further evidence for the existence of extraterrestrial beings who may have visited Earth and influenced human development.

The implications of these ancient stories and their potential connections to extraterrestrial life extend beyond our understanding of human history. As we continue to explore the cosmos, search for habitable planets, and seek evidence of extraterrestrial life, the Ancient Alien Theory serves as a reminder that we may not be alone in the universe. By examining these ancient narratives, we can gain insights into the nature of potential extraterrestrial civilizations, their motivations for visiting Earth, and the ways in which they may have influenced our development.

In conclusion, the Ancient Alien Theory provides a compelling framework for interpreting the various myths, legends, and stories of ancient alien environments and species. By examining these narratives in the context of this theory, we can gain a greater appreciation for the potential existence of advanced extraterrestrial civilizations and their interactions with humanity throughout history. As we continue to explore our own place in the cosmos, these ancient narratives serve as a reminder that we may not be alone in the universe and that the stars may hold the key to unlocking the mysteries of our past.

Chapter 16

Theories About Alien Life, Culture and Interactions

Worlds Outside Our Own

The building blocks of extraterrestrial life could potentially differ from those found on Earth. While terrestrial life relies primarily on carbon-based molecules and water, alien life forms might utilize alternative biochemistries based on elements such as silicon or ammonia. The discovery of organisms on Earth that can metabolize arsenic, for example, demonstrates the potential for life to develop using a variety of chemical elements. This flexibility in the building blocks of life expands the range of potential environments in which extraterrestrial life could thrive and supports the idea that ancient aliens could have existed in diverse forms.

The adaptability of life on Earth provides a strong indication that extraterrestrial life could exist in a wide range of environments. Terrestrial organisms have evolved to inhabit various ecosystems, from the depths of the ocean to the frigid polar regions. This adaptability suggests that alien life could similarly evolve to survive in diverse environments, such as the surface of icy moons, the high-pressure depths of gas giants, or the radiation-bathed surfaces of distant exoplanets. The potential for life to adapt to diverse conditions lends credence to the idea that ancient aliens could have inhabited various worlds and interacted with early human civilizations.

The concept of speciation, the process by which new species arise through evolution, supports the idea of a diverse range of extraterrestrial life forms. The multitude of species that have evolved on Earth, each with its unique traits and characteristics, demonstrates the potential for similarly diverse life to evolve elsewhere in the universe. Ancient aliens might have developed along unique evolutionary pathways, giving rise to distinct biological structures, sensory capabilities, and cognitive abilities. This diversity in form and function could have significantly influenced the interactions between ancient aliens and early human civilizations, shaping the ways in which knowledge and technology were exchanged.

The potential morphological and physiological diversity of extraterrestrial life forms is a fascinating aspect of Ancient Alien Theory. Alien life could exhibit a vast array of physical forms, ranging from microscopic organisms to colossal, multicellular entities. These life forms might also possess unique physiological adaptations, such as the ability to withstand extreme temperatures, resist radiation, or metabolize unconventional energy sources. The morphological and physiological diversity of ancient aliens could have played a crucial role in their interactions with early human civilizations, with their unique abilities contributing to the development of advanced technologies and the shaping of human history.

The incredible diversity of potential extraterrestrial life forms has profound implications for Ancient Alien Theory. The existence of such diverse life in the universe increases the likelihood that ancient human civilizations encountered otherworldly beings with unique abilities and knowledge. These encounters could have shaped the course of human history, inspiring myths, legends, and cultural beliefs that persist to this day. The possibility of diverse extraterrestrial life forms lends credence to the idea that ancient aliens not only existed but played a significant role in the development of human civilization.

The Drake Equation, while inherently speculative, offers a compelling framework for estimating the number of civilizations in the Milky Way that could potentially engage in communication with Earth. By considering a range of factors, the equation allows researchers to weigh the probabilities of various aspects of cosmic evolution and planetary development. As our understanding of these factors improves through advancements in astronomy and astrobiology, the Drake Equation becomes an increasingly valuable tool in the search for extraterrestrial intelligence.

The Ancient Alien Theory proposes that advanced extraterrestrial civilizations have visited Earth in the past, influencing human development and culture. When considered in the context of the Drake Equation, this theory suggests that the development of intelligent life may be more common than previously thought, and that the likelihood of contact between civilizations may be higher. If the factors in the Drake Equation are adjusted to account for the possibility of ancient alien interactions, the estimated number of communicative civilizations in the galaxy could increase, providing further support for the existence of ancient aliens.

In recent years, the discovery of exoplanets orbiting stars outside our solar system has expanded our knowledge of planetary systems and habitable environments. This information directly impacts the factors in the Drake Equation, such as the fraction of stars with planetary systems and the number of planets that could potentially support life. As more exoplanets are discovered and analyzed, the probability of finding planets with conditions suitable for the development of intelligent life increases, lending credence to the idea that ancient aliens may have existed and visited Earth.

As our understanding of the universe continues to grow, the assumptions and factors in the Drake Equation can be refined and updated to more accurately reflect our current knowledge. For example, recent discoveries about extremophiles, organisms capable of surviving in extreme environments, can influence our understanding of habitable zones around stars and the likelihood of life developing on planets with different conditions. By continually revisiting the Drake Equation with new data and insights, researchers can maintain an up-to-date and informed perspective on the probability of intelligent life in the universe and its potential interactions with Earth.

The Drake Equation has played a significant role in justifying the search for extraterrestrial intelligence (SETI) and guiding research in the field. SETI efforts focus on detecting signals from advanced civilizations that may be trying to communicate with other intelligent beings in the galaxy. By providing a framework for estimating the number of such civilizations, the Drake Equation helps to support the underlying premise of SETI: that the existence of intelligent life elsewhere in the galaxy is not only possible, but likely. As a result, the Drake Equation and the Ancient Alien Theory can be seen as complementary ideas, both contributing to our understanding of the potential for extraterrestrial life and its implications for human history.

Astrobiologists studying extremophiles on Earth have discovered an incredible array of organisms capable of surviving in environments previously thought to be inhospitable to life. These findings have significant implications for the search for extraterrestrial life, as they suggest that life might exist in a variety of forms and thrive in conditions that differ greatly from those found on our planet. The discovery of extremophiles strengthens the argument for ancient alien existence, as it broadens the range of potential habitats for life beyond Earth and supports the idea that alien life forms could have visited our planet in the past.

The study of exoplanets, or planets orbiting stars outside our solar system, has revealed numerous potentially habitable worlds. These planets, located within the "habitable zone" of their respective stars, have conditions that may be conducive to the development of life. Astrobiologists analyze the chemical composition and atmospheric conditions of these exoplanets to determine their suitability for life. The discovery of such worlds bolsters the Ancient Alien Theory by increasing the likelihood that life exists elsewhere in the universe, and that advanced extraterrestrial civilizations may have had the opportunity to interact with humanity throughout history.

Astrobiologists also focus their efforts on investigating the potential for microbial life on celestial bodies within our solar system, such as Mars and Europa. Mars, in particular, has been the subject of intense study due to the presence of liquid water beneath its surface and the detection of organic molecules. Similarly, Jupiter's moon Europa is believed to harbor a subsurface ocean that may provide the necessary conditions for life to emerge. The possibility of microbial life on these celestial bodies strengthens the argument for ancient alien interactions with Earth, as it demonstrates that life can exist in a wide range of environments and increases the likelihood that advanced extraterrestrial civilizations could have emerged elsewhere in the universe.

Another area of astrobiology that supports the Ancient Alien Theory is the study of panspermia, the hypothesis that life on Earth may have originated from elsewhere in the universe. Panspermia proposes that microbial life could have been transported to Earth via comets, meteorites, or even by ancient alien visitors. This theory challenges the traditional view of abiogenesis, or the spontaneous emergence of life from non-living matter, and suggests that life on Earth may be intimately connected to the broader cosmos. If panspermia is proven to be a valid process, it would significantly increase the likelihood of ancient alien interactions with Earth, as it would demonstrate a direct link between extraterrestrial life and the development of life on our planet.

One possible solution to the Fermi Paradox, which aligns with the Ancient Alien Theory, is the Zoo Hypothesis. This idea proposes that extraterrestrial civilizations might be observing humanity without interfering, much like humans observe animals in a zoo. This concept is similar to the idea of the "Prime Directive" in the science fiction series Star Trek, which forbids interference with the natural development of less advanced civilizations. If this hypothesis is accurate, advanced alien civilizations may have interacted with us in the past but now choose to remain hidden, observing our progress from a distance.

The Ancient Alien Theory proposes that there is substantial evidence of past interactions between extraterrestrial beings and human civilizations. This evidence includes megalithic structures, like the Great Pyramid of Giza and the Nazca Lines, which appear to be constructed using advanced technologies beyond the capabilities of ancient humans. Additionally, ancient texts and mythologies from various cultures around the world contain stories of encounters with otherworldly beings and celestial events. These accounts could be interpreted as evidence of extraterrestrial visitation and involvement in human history, thus providing an answer to the Fermi Paradox.

Another aspect of the Fermi Paradox is the possibility that intelligent civilizations might inevitably self-destruct before they have a chance to make contact with other civilizations. In this scenario, the Ancient Alien Theory could suggest that extraterrestrial beings intervened in human history to prevent our self-destruction or to help us advance more rapidly than we would have on our own. This intervention might have taken the form of technological assistance, the sharing of knowledge, or even direct guidance in the development of human societies.

The Great Filter is a concept related to the Fermi Paradox that suggests there may be a critical barrier that prevents intelligent civilizations from reaching a level of technological advancement that allows for interstellar communication or travel. This filter could be a natural barrier, like a cataclysmic event, or a self-imposed one, such as a species' tendency towards self-destruction. The Ancient Alien Theory offers an alternative perspective, proposing that extraterrestrial beings may have assisted humanity in overcoming this filter by providing us with the knowledge and technology necessary to advance more rapidly than we would have on our own.

The Fermi Paradox raises the question of why we have not yet encountered evidence of extraterrestrial life if it is so prevalent in the universe. The Ancient Alien Theory suggests that our perception of reality might be influenced by the interactions we have had with extraterrestrial beings in the past. These beings could have shaped our beliefs, myths, and cultural narratives, thus affecting the way we view the cosmos and our place in it. In this context, the Fermi Paradox might be less of a paradox and more a reflection of our limited understanding of the universe and the role of extraterrestrial life in our history.

The ancient Sumerian civilization, considered to be one of the earliest in human history, left behind numerous texts describing the Anunnaki, a group of gods believed to have descended from the sky. Ancient Alien Theory proponents argue that these accounts are not merely mythological, but rather evidence of extraterrestrial beings who visited Earth and played a crucial role in the development of Sumerian society. The Anunnaki are often credited with bestowing knowledge, technology, and advanced agricultural techniques upon the Sumerians, thus helping to advance their civilization.

Ancient Egyptian mythology is filled with references to gods associated with celestial bodies, such as the sun god Ra and the moon god Thoth. Proponents of Ancient Alien Theory believe that these deities may have been inspired by extraterrestrial visitors who interacted with the Egyptians, sharing their knowledge and technology to help shape this advanced civilization. This idea is supported by the remarkable architectural achievements of the Egyptians, such as the construction of the Great Pyramid of Giza, which is believed to incorporate advanced mathematical and astronomical knowledge that would have been challenging to achieve without external assistance.

The Indian Vedas and other ancient Sanskrit texts contain numerous references to flying machines called Vimanas, which were allegedly used by the gods to travel through the sky. Ancient Alien Theory proponents argue that these accounts may represent a record of advanced extraterrestrial technology, used by ancient alien visitors to travel across Earth and interact with early human civilizations. These ancient texts often describe the Vimanas as being equipped with advanced weaponry and powered by complex propulsion systems, further supporting the idea that their origins may lie beyond Earth.

Another aspect of Ancient Alien Theory is the idea that extraterrestrial beings may have played a role in the genetic development of humanity. Proponents of this theory suggest that ancient aliens may have manipulated human DNA to accelerate our evolution or to create hybrid species. This idea is supported by various ancient myths and legends that describe gods or celestial beings engaging in the creation of humans or interbreeding with them. The potential for genetic manipulation raises fascinating questions about the origins of human intelligence, physical capabilities, and even our diverse cultural traditions.

The notion that ancient alien visitors interacted with early human civilizations is not limited to specific regions or cultures. Evidence supporting Ancient Alien Theory can be found across the globe, from the Mayan civilization in Central America to the ancient Chinese culture in East

Asia. These widespread accounts of otherworldly beings suggest that extraterrestrial involvement in human development was not an isolated phenomenon, but rather a global phenomenon that played a crucial role in shaping the course of human history.

In addition to the story of Prometheus, Greek mythology contains numerous examples that could be interpreted as evidence of ancient alien contact. For instance, the gods of Olympus were said to possess advanced knowledge and wield powerful weapons, such as Zeus's thunderbolts, which might be viewed as an ancient description of advanced technology. Moreover, the demigods, who were the offspring of human and divine unions, can be seen as potential examples of ancient alien genetic manipulation or hybridization.

Native American cultures across the continent have shared stories of the Star People, who are often described as celestial beings that visited Earth, imparting wisdom and knowledge to their ancestors. The Hopi people of the American Southwest, for example, believe in the Kachinas, otherworldly beings who descended from the sky and brought teachings that were essential for their survival and cultural development. These stories of the Star People not only demonstrate the prevalence of ancient alien themes in Native American mythology but also suggest that these encounters had a significant impact on their societies.

The Sumerian and Mesopotamian civilizations are also rich sources of myths and legends that can be interpreted as evidence of ancient alien contact. The Anunnaki, a group of gods mentioned in Sumerian texts, were said to have come from the heavens and bestowed knowledge and technology upon humanity. The Epic of Gilgamesh, one of the earliest known literary works, contains references to gods with otherworldly powers and advanced technologies, such as the ability to fly and control the elements. These accounts provide compelling support for the idea that ancient aliens played a role in shaping the beliefs and cultural development of these early civilizations.

Mesoamerican mythology, encompassing the beliefs of the Maya, Aztec, and other ancient cultures in Central America, also contains stories that could be interpreted as evidence of ancient alien interactions. The Mayan deity Kukulkan, also known as Quetzalcoatl among the Aztecs, was described as a feathered serpent who descended from the heavens and provided humanity with advanced knowledge and technologies. The Mayan calendar and complex astronomical knowledge, which were far ahead of their time, could be seen as the result of extraterrestrial influence and guidance.

Chinese mythology also contains elements that suggest the possibility of ancient alien contact. One of the most significant examples is the story of the Yellow Emperor, a legendary figure credited with founding Chinese civilization. According to legend, the Yellow Emperor was born after his mother was visited by a divine being who descended from the sky in a "thunderclap." His rule was marked by the introduction of advanced technologies, such as the compass, silk production, and the wheel, which could be seen as evidence of extraterrestrial influence on Chinese culture and development.

The influence of ancient aliens on human mythology and cultural development is evident across numerous cultures and civilizations. The various stories and legends that describe otherworldly beings, advanced technologies, and celestial events serve as a testament to the profound impact that these encounters may have had on the beliefs and societies of our ancestors. As we continue to explore the depths of our past, the possibility of ancient alien contact remains an intriguing and thought-provoking aspect of our understanding of human history.

The possibility that ancient aliens used telepathy or other advanced methods of communication to interact with humans is a fascinating aspect of Ancient Alien Theory. Telepathic communication would enable extraterrestrial beings to bypass the constraints of spoken language and directly convey thoughts, ideas, and knowledge to their human counterparts. This form of communication might also explain the prevalence of visions and prophetic dreams in ancient texts, which could be interpreted as telepathic transmissions from advanced beings.

Another possibility is that ancient aliens learned to speak human languages, allowing them to communicate more directly with early civilizations. This theory suggests that extraterrestrial beings had the capacity to quickly learn and adapt to the linguistic systems of various human cultures. This mastery of language would have allowed them to share their knowledge and technology more effectively, contributing to the rapid development of early human societies.

Ancient Alien Theory also proposes that extraterrestrial beings may have influenced the development of early written languages, such as cuneiform and hieroglyphics. By introducing new systems of writing or refining existing ones, ancient aliens could have facilitated the spread of knowledge and ideas across different cultures. This influence on written language might be evident in the similarities between ancient scripts, which could be attributed to a common extraterrestrial source.

Nonverbal communication and the use of symbolism could have played a significant role in ancient alien interactions with humans. Extraterrestrial beings may have used symbols, gestures, or visual representations to convey complex ideas and concepts. This form of communication would transcend linguistic barriers, enabling ancient humans to understand and interpret the messages being conveyed. The prevalence of similar symbols and motifs across different cultures and time periods could be seen as evidence of a shared extraterrestrial influence.

In order to effectively communicate with early human civilizations, ancient aliens may have needed to adapt to the cultural norms and practices of these societies. This adaptation could involve adopting the appearance of gods or other revered figures, as well as engaging in rituals and ceremonies that held significance for the humans they encountered. By assimilating into human culture, ancient aliens would have been better equipped to share their knowledge and technology, while also maintaining a position of authority and respect.

The influence of ancient aliens on human language and communication could have far-reaching implications for our understanding of history and the development of culture. If extraterrestrial beings did indeed interact with early human societies, their influence on language and communication might still be felt today. This impact could manifest in the form of linguistic similarities between unrelated languages, as well as the shared use of symbols and motifs that can be traced back to ancient alien encounters. By considering the potential role of ancient aliens in the evolution of human language and communication, we gain a deeper understanding of our own history and our place in the universe.

The Indian Vedas, ancient Sanskrit texts, contain numerous references to flying machines called Vimanas. These descriptions often include remarkable details of their propulsion systems, aerial maneuvers, and even weaponry. Proponents of Ancient Alien Theory argue that these accounts are evidence of advanced extraterrestrial technology that was shared with or observed by ancient humans. The knowledge of such advanced technology could have spurred the development of early aviation and inspired countless myths and legends about gods traveling through the skies.

The construction of the Great Pyramid of Giza is often cited as evidence of extraterrestrial influence on human development. Despite its age, the pyramid exhibits incredible precision in its alignment, dimensions, and internal structure. Theories suggest that the builders of the pyramid employed advanced engineering techniques, possibly introduced by ancient aliens, to create this architectural marvel. The knowledge of such advanced techniques could have had a profound impact on the development of human engineering and construction.

Puma Punku, an ancient archaeological site in Bolivia, is home to some of the most intricate and precise stonework found anywhere in the world. The site features massive stone blocks that have been cut and shaped with incredible accuracy, fitting together seamlessly without the use of mortar. Proponents of Ancient Alien Theory argue that this level of craftsmanship would have been impossible for the ancient people of the region to achieve without the aid of advanced extraterrestrial technology. The sharing of such knowledge could have had significant implications for the development of human construction techniques and architectural design.

The Antikythera Mechanism, an ancient Greek artifact discovered in 1901, is often cited as evidence of advanced technology in antiquity. This complex mechanical device, which predates any known similar technology by more than a millennium, is believed to have been used for astronomical calculations and predictions. Ancient Alien Theory proponents argue that the knowledge required to create such a sophisticated device must have come from an advanced extraterrestrial civilization. This exchange of knowledge could have had far-reaching consequences for the development of human technology, particularly in the fields of mathematics, astronomy, and engineering.

The Nazca Lines, a series of enormous geoglyphs etched into the desert in southern Peru, have long been a source of fascination and debate among researchers. These massive drawings, which can only be fully appreciated from the air, depict various geometric shapes, animals, and other figures. Ancient Alien Theory proponents argue that the creation of these geoglyphs required advanced technology and knowledge, possibly provided by extraterrestrial visitors. The presence of such large-scale, aerially-visible artwork could suggest an attempt to communicate with or pay tribute to these ancient alien beings, further demonstrating their influence on human development.

The potential impact of ancient alien interactions on human religious practices is a key component of the Ancient Alien Theory. The presence of extraterrestrial beings may have been perceived as divine, leading to their incorporation into various religious beliefs and rituals. For instance, the Sumerian gods, known as the Anunnaki, are often associated with the planets and stars, suggesting a possible extraterrestrial origin. Similarly, the ancient Egyptian pantheon includes numerous gods associated with celestial bodies, such as Ra, the sun god, and Thoth, the moon god. These examples demonstrate how ancient alien visitors may have shaped human religious practices by being interpreted as divine entities.

The presence of ancient aliens could also have had significant implications for the development of political systems and leadership structures in early human civilizations. Extraterrestrial beings with advanced knowledge and technology may have been seen as powerful and wise, leading humans to seek their guidance in matters of governance and decision-making. For example, the concept of divine right, which asserts that a ruler's authority comes directly from a higher power, could have been influenced by the presence of ancient aliens who were perceived as gods. This would have given these beings considerable influence over the political landscape of the time.

The interactions between ancient aliens and human societies may have also had profound effects on cultural norms and social structures. The exchange of knowledge, technology, and ideas between humans and extraterrestrial beings could have led to the development of new customs, values, and ways of life. For instance, the ancient Greek story of

Prometheus, who stole fire from the gods and gave it to humanity, can be seen as a metaphor for the transfer of advanced knowledge from extraterrestrial visitors to humans. This knowledge could have had a transformative effect on human society, shaping cultural norms and social structures in ways that still influence our world today.

Another aspect of the socio-political impact of ancient alien interactions is their potential involvement in warfare and conflict. Ancient texts and myths often describe epic battles between gods, which could be interpreted as accounts of extraterrestrial beings engaging in combat or influencing human conflicts. The ancient Indian epic, the Mahabharata, for example, contains descriptions of advanced weapons and flying machines that could be interpreted as evidence of extraterrestrial technology. If ancient aliens did play a role in human warfare, their advanced knowledge and technology could have significantly impacted the outcomes of these conflicts and shaped the course of human history.

The socio-political impacts of ancient alien interactions continue to resonate in our modern world. The cultural, religious, and political systems that were influenced by these beings have left a lasting mark on human civilization. As we continue to explore our past and seek answers to the mysteries of our origins, the possibility of ancient alien involvement in the development of human society remains a compelling and thought-provoking topic for researchers, historians, and enthusiasts alike.

One perspective within Ancient Alien Theory proposes that extraterrestrial visitors were benevolent beings who sought to guide humanity toward greater knowledge and understanding. This viewpoint is supported by the numerous accounts of ancient aliens providing advanced technology, knowledge, and even genetic enhancement to early human civilizations. Examples include the Sumerian tales of the Anunnaki, who supposedly shared their knowledge of agriculture, writing, and construction with humanity, and the Egyptian god Thoth, who is credited with introducing writing, mathematics, and other intellectual pursuits to ancient Egypt. These ancient alien encounters can be seen as positive collaborations that allowed human societies to progress rapidly and achieve remarkable advancements.

Another perspective within Ancient Alien Theory proposes that extraterrestrial beings had more selfish motives for interacting with early human civilizations, such as resource extraction or the use of humans for genetic experimentation. This viewpoint suggests that ancient aliens could have been interested in Earth's resources, such as gold, minerals, or water, and may have enlisted human labor for their extraction. Additionally, proponents of this theory argue that ancient aliens might have manipulated human genetics to create hybrid species, improve human capabilities, or even control human populations. Evidence for this perspective can be found in myths and legends that describe encounters with otherworldly beings who demanded tribute, resources, or labor from human societies.

It is important to consider that the intentions and motivations of ancient alien visitors might not be easily categorized as either benevolent or malevolent. The myriad of ancient stories and legends that describe encounters with otherworldly beings often presents a complex and nuanced picture of their intentions and the consequences of their actions. Some interactions may have been mutually beneficial, with both human and extraterrestrial civilizations gaining from the exchange of knowledge and resources. Other encounters may have involved manipulation, coercion, or even conflict, as ancient aliens pursued their own objectives at the expense of human societies.

The ethical considerations and motivations of extraterrestrial visitors can be evaluated by examining the ancient accounts, myths, and legends that describe their interactions with early human civilizations. These stories can provide valuable insights into the nature of these encounters, the intentions of the extraterrestrial beings, and the impact of their actions on human societies. By analyzing these accounts, we can gain a better understanding of the possible ethical dimensions of ancient alien interactions with humanity and how these encounters may have shaped our history and our understanding of the cosmos.

Ancient Alien Theory raises intriguing questions about the ethical considerations and motivations of extraterrestrial visitors who may have interacted with early human civilizations. Examining the variety of ancient stories and legends that describe encounters with otherworldly beings can provide valuable insights into the intentions and consequences of these interactions, offering a more nuanced and comprehensive understanding of the potential impact of ancient aliens on human history and development.

As the search for exoplanets accelerates, we continue to discover new worlds with a diverse range of characteristics. This expanding catalog of planets outside our solar system provides fertile ground for researchers who favor the existence of ancient aliens. The discovery of Earth-like planets orbiting within their stars' habitable zones could provide potential locations where extraterrestrial life evolved and thrived. As we continue to study these worlds, we may uncover signs of past or present life, lending further credence to the Ancient Alien Theory.

The exploration of celestial bodies within our solar system, such as Mars, Europa, and Enceladus, has revealed the potential for subsurface liquid water and other conditions that could potentially support life. Scientists favoring the Ancient Alien Theory may argue that life could have originated in these environments and later traveled to Earth through panspermia, a process where life is transferred between planets via meteorites or other celestial bodies. The discovery of microbial life on these nearby celestial bodies could provide a basis for understanding the potential interactions between ancient aliens and early human civilizations.

The search for extraterrestrial intelligence (SETI) is an ongoing effort to detect signals from alien civilizations. As technology continues to advance, our ability to detect potential signals from extraterrestrial sources improves. Proponents of the Ancient Alien Theory argue that the discovery of a signal from an intelligent alien civilization could validate the idea that ancient aliens visited Earth and interacted with early human societies. As we develop more sophisticated tools to listen for and interpret these signals, the possibility of contact with other intelligent beings becomes more plausible.

Artificial intelligence and machine learning have the potential to revolutionize our search for extraterrestrial life and our understanding of the Ancient Alien Theory. By employing advanced algorithms and pattern recognition techniques, researchers can analyze vast quantities of data from astronomical observations, potentially uncovering new insights into the existence of ancient alien civilizations. For instance, AI-driven analysis could help identify potential artificial structures on distant planets or detect patterns in extraterrestrial signals that would otherwise be missed by human researchers.

The study of ancient alien interactions with humanity may benefit from the integration of multiple disciplines, including archaeology, anthropology, astronomy, and astrobiology. By combining the expertise of researchers from these diverse fields, we can develop a more comprehensive understanding of the potential evidence for ancient alien visitation. As new discoveries are made, and our understanding of the cosmos deepens, these interdisciplinary approaches may shed new light on the impact of ancient alien encounters on human history and culture.

As we continue to explore the cosmos and search for evidence of extraterrestrial life, the Ancient Alien Theory has the potential to reshape our perception of humanity's place in the universe. If evidence emerges that supports the idea that ancient aliens visited Earth and influenced human development, it could challenge our understanding of human history and encourage us to rethink our assumptions about our own capabilities and potential. The possibility of ancient alien interactions with humanity invites us to consider a future where we too may explore the stars and interact with other intelligent life forms, continuing the legacy of our ancient alien ancestors.

Chapter 17

Theories of Intergalactic Travel

Can We One Day Travel to the Stars?

The allure of intergalactic travel has inspired generations of scientists, researchers, and science fiction writers to imagine the possibility of humanity exploring the cosmos, encountering other civilizations, and establishing contact with extraterrestrial beings. Proponents of the Ancient Alien Theory contend that evidence of past intergalactic travel can be found in the form of ancient artifacts, structures, and texts that point to the presence of advanced knowledge and technologies beyond the capabilities of the civilizations of that time. In this context, the theories and scientific advancements that could make intergalactic travel possible lend credibility to the idea that ancient aliens may have visited Earth in the past.

The Ancient Alien Theory suggests that extraterrestrial beings may have played a role in shaping human history by sharing their advanced knowledge and technology with ancient civilizations. This hypothesis is supported by the presence of enigmatic structures, such as the pyramids of Egypt and the Nazca Lines in Peru, which display remarkable engineering feats that seemingly defy explanation by the technology available to those ancient societies.

Additionally, ancient texts and myths from various cultures around the world, including the Sumerian cuneiform tablets and the Indian epics like the Mahabharata and the Ramayana, describe encounters with celestial beings and the use of advanced technologies. These accounts, according to proponents of the Ancient Alien Theory, provide further evidence that intergalactic travelers may have visited Earth and influenced our ancestors.

To support the possibility of ancient alien encounters, it is essential to examine the scientific advancements that could enable intergalactic travel. The development of advanced propulsion systems, the manipulation of spacetime, and the potential use of artificial intelligence and robotics in space exploration are all areas of study that may one day make intergalactic travel a reality. By understanding the cutting-edge research in these fields, we can better appreciate the plausibility of ancient alien visitations and their potential impact on human history.

Furthermore, the ongoing search for extraterrestrial life and the exploration of exoplanets in the habitable zones around other stars contribute to the possibility of intergalactic travel and the existence of ancient aliens. The discovery of microbial life on Mars or the detection of atmospheric biosignatures on exoplanets could provide tantalizing hints that life exists beyond Earth, further supporting the idea that advanced extraterrestrial civilizations may have been capable of intergalactic travel and have visited our planet in the past.

The possibility of intergalactic travel has far-reaching implications for our understanding of human history and the potential existence of ancient aliens. By examining the theories and scientific advancements that could make intergalactic travel possible, we can better appreciate the likelihood that ancient aliens may have visited Earth and influenced our ancestors. As we continue to push the boundaries of our understanding of the cosmos, the Ancient Alien Theory will remain an intriguing and thought-provoking concept that challenges conventional explanations and invites us to consider the mysteries of our past.

The immense distances involved in intergalactic travel represent one of the main challenges in reaching other galaxies. With our nearest neighboring galaxy, Andromeda, located approximately 2.537 million light-years away, it would take billions of years to reach it using our fastest spacecraft. As such, the idea that ancient aliens visited Earth seems improbable when considering the vast distances they would have had to travel. However, proponents of Ancient Alien Theory argue that advanced extraterrestrial civilizations may have overcome these challenges through innovative methods and technologies that allowed them to traverse vast interstellar distances within a human lifetime.

One way that ancient aliens might have achieved this feat is by developing propulsion systems far more advanced than anything humanity has yet to conceive. Such technology would allow them to cover the enormous distances between galaxies at speeds that would make intergalactic travel possible within a reasonable time frame. Some of these proposed methods include bending space-time, using wormholes, or harnessing the energy of stars for propulsion.

Another possibility is that ancient aliens were able to manipulate time itself, effectively reducing the duration of their journey. By altering their perception of time or even slowing down time relative to their spacecraft, they could have traversed the vast distances between galaxies without experiencing the same passage of time as we understand it. This concept, while purely speculative, would provide a potential solution to the challenge of intergalactic travel and lend further credence to the Ancient Alien Theory.

Additionally, proponents of Ancient Alien Theory suggest that these advanced civilizations may have possessed a more profound understanding of the universe's fundamental nature, allowing them to overcome the limitations of our current understanding of physics. Through this knowledge, ancient aliens might have discovered methods of travel that seem impossible from our current scientific perspective, such as teleportation or other forms of instantaneous travel.

Ultimately, the challenges of intergalactic travel, while daunting, do not entirely discount the possibility of ancient alien visitations. By considering the potential advancements and technologies that advanced extraterrestrial civilizations might have developed, we can begin to appreciate the plausibility of ancient aliens overcoming these challenges and influencing human history. As our understanding of the universe continues to expand, we may one day uncover the secrets of intergalactic travel, providing further evidence in support of the Ancient Alien Theory.

One proposed method for achieving faster-than-light travel involves the hypothetical existence of tachyons, particles that are thought to travel faster than the speed of light. While tachyons have not been detected or proven to exist, their theoretical properties make them an intriguing possibility for interstellar travel. Proponents of Ancient Alien Theory argue that extraterrestrial civilizations may have harnessed the power of tachyons or other superluminal particles to traverse vast distances, visiting Earth and influencing ancient cultures.

Quantum entanglement is a phenomenon in which the quantum states of two or more particles become linked, resulting in instantaneous changes to one particle's state when the other is altered, regardless of the distance between them. This phenomenon has led to the concept of "quantum teleportation," which, although not yet possible for macroscopic objects, raises the possibility of instantaneous communication over vast interstellar distances. Ancient Alien Theory proponents suggest that advanced extraterrestrial civilizations may have developed technology based on quantum entanglement, allowing them to communicate and potentially travel across the cosmos with ease.

Another proposed method for circumventing the light-speed barrier is the concept of travel through alternative dimensions or hyperspace. This idea involves the existence of additional spatial dimensions beyond our familiar three-dimensional universe, which could be accessed and utilized for rapid interstellar travel. While the existence of alternative dimensions remains purely theoretical, the concept has been explored in various scientific disciplines, including string theory and M-theory.

Supporters of the Ancient Alien Theory argue that ancient artifacts and texts describing otherworldly beings and their advanced technologies could be evidence of extraterrestrial visitors who mastered the ability to travel through alternative dimensions or hyperspace. These advanced civilizations may have shared their knowledge with ancient humans, explaining the seemingly advanced understanding of the cosmos and engineering found in various ancient cultures.

Negative mass, a hypothetical form of matter with mass opposite to that of ordinary matter, has been proposed as another potential means of achieving faster-than-light travel. Negative mass would exhibit properties such as repelling rather than attracting other matter and accelerating in the opposite direction of applied force. If harnessed, negative mass could potentially enable faster-than-light travel by counteracting the relativistic mass increase experienced by objects approaching the speed of light.

Proponents of the Ancient Alien Theory speculate that advanced extraterrestrial civilizations may have discovered and utilized negative mass or other exotic forms of matter to enable interstellar travel, visiting Earth and influencing the development of human history. The existence of such exotic matter, combined with ancient accounts of advanced technologies and seemingly impossible construction methods, lends credibility to the idea that ancient aliens may have possessed the ability to travel faster than light.

Wormholes and the concept of instantaneous travel through spacetime have been featured in numerous ancient texts and mythologies, often associated with supernatural beings or advanced technologies. Proponents of the Ancient Alien Theory argue that these accounts may be evidence of extraterrestrial beings utilizing wormholes to visit Earth in the distant past.

For example, in Hindu mythology, the concept of the "Vimana" – a flying chariot or palace – is often described as having the ability to travel vast distances in the blink of an eye. Similarly, ancient Greek and Roman mythologies describe gods and demigods with the power to travel instantly from one location to another. These accounts, according to Ancient Alien Theory proponents, may represent encounters with extraterrestrial beings who used wormhole technology to visit Earth.

One of the main challenges in realizing wormhole-based interstellar travel is the stabilization of these theoretical structures. According to current understanding, a stable wormhole would require a form of "exotic matter" with negative energy density to prevent the wormhole from collapsing. This exotic matter has yet to be discovered or synthesized, but its existence has been proposed by some physicists as a potential solution to the stabilization problem.

Ancient Alien Theory proponents argue that advanced extraterrestrial civilizations may have already discovered or developed the necessary exotic matter to create stable wormholes, allowing them to visit Earth and influence ancient cultures. This idea is supported by the presence of seemingly impossible construction methods and advanced knowledge in ancient civilizations, which could be the result of contact with technologically advanced extraterrestrial visitors.

While the creation and stabilization of artificial wormholes remain a significant challenge, some scientists have suggested that naturally occurring wormholes may exist in the universe. These natural wormholes could be formed by various cosmic processes, such as the interaction of massive celestial bodies or as a result of cosmic inflation in the early universe.

If naturally occurring wormholes exist, they could potentially be harnessed by advanced extraterrestrial civilizations for interstellar travel. Proponents of the Ancient Alien Theory argue that the existence of natural wormholes could explain the presence of advanced knowledge and technology in ancient cultures, as well as the accounts of encounters with otherworldly beings throughout human history.

Some proponents of the Ancient Alien Theory speculate that advanced extraterrestrial civilizations may have established vast, wormhole-based communication and transportation networks throughout the cosmos. These networks could enable instantaneous travel and communication between distant star systems and galaxies, allowing these civilizations to maintain contact with one another and explore the far reaches of the universe.

This idea is supported by the numerous accounts of ancient civilizations receiving knowledge and guidance from celestial beings or gods, often through mysterious means or devices. According to Ancient Alien Theory proponents, these accounts could represent encounters with extraterrestrial visitors who utilized wormhole-based communication and transportation networks to share their advanced knowledge and technology with ancient human societies.

The Alcubierre Drive has garnered significant attention in both the scientific community and popular culture due to its potential to revolutionize space travel. The concept of warping spacetime to enable faster-than-light travel presents a solution to the vast distances that separate celestial bodies, allowing for more efficient exploration of the cosmos. This theoretical breakthrough, while yet to be realized, has given rise to exciting possibilities that may one day validate the existence of ancient aliens.

Supporters of the Ancient Alien Theory argue that the Alcubierre Drive might be the key to understanding how extraterrestrial civilizations could have visited Earth in the past. They believe that such advanced beings may have mastered the technology to manipulate spacetime and traverse interstellar distances with relative ease.

Proponents of the Ancient Alien Theory often point to the sophisticated knowledge and seemingly inexplicable construction techniques found in various ancient civilizations as evidence of extraterrestrial influence. They argue that the Alcubierre Drive technology could have been shared with our ancestors, allowing them to achieve incredible feats that remain baffling to modern researchers.

Accounts of encounters with extraterrestrial beings throughout human history could potentially be explained by the existence of the Alcubierre Drive. If extraterrestrial civilizations had access to such technology, they would have been able to visit Earth and interact with human societies without violating the laws of physics. This could account for the numerous myths, legends, and historical accounts of otherworldly beings with advanced knowledge and capabilities.

As our understanding of physics and the nature of spacetime continues to evolve, the Alcubierre Drive may become more than just a theoretical concept. If humanity were to develop this technology, it could provide irrefutable evidence that intergalactic travel is possible, lending credibility to the idea that ancient aliens once visited Earth.

The development of the Alcubierre Drive could also open up new avenues of research and exploration, allowing us to investigate the possibility of ancient alien encounters more thoroughly. In turn, this could lead to a greater understanding of our own history and the potential existence of extraterrestrial civilizations throughout the universe.

The Alcubierre Drive represents an exciting and promising concept in the realm of intergalactic travel. While still purely theoretical, its potential applications and implications for both the Ancient Alien Theory and our understanding of the cosmos make it a fascinating area of study. As we continue to delve deeper into the mysteries of spacetime and the nature of the universe, the Alcubierre Drive may one day provide the key to unlocking the secrets of ancient alien visitations and the vast potential of interstellar exploration.

Antimatter propulsion holds great promise for intergalactic travel due to the tremendous amount of energy released during the annihilation process. To harness this energy, a spacecraft would need to store and contain antimatter safely, as even minute amounts of antimatter reacting with matter would result in catastrophic explosions. Ancient Alien Theory proponents argue that advanced extraterrestrial civilizations might have developed efficient methods for producing, storing, and utilizing antimatter for propulsion, giving them the capability to travel vast distances across the cosmos.

Another exotic energy source that could potentially be harnessed for intergalactic travel is zero-point energy. According to quantum field theory, even in the absence of particles and at the lowest possible energy state, there are still energy fluctuations in the vacuum. This energy, known as zero-point energy, is theorized to be virtually unlimited. If a civilization could tap into and harness this energy, it would have access to an immense power source capable of fueling intergalactic voyages.

Proponents of Ancient Alien Theory suggest that extraterrestrial beings might have already developed the technology to exploit zero-point energy, enabling them to visit Earth and share their knowledge with ancient civilizations. This could explain the presence of advanced technologies and complex structures in the archaeological record that seem to defy conventional explanations.

Fusion power, the process by which atomic nuclei are combined to release energy, has long been considered the holy grail of energy production. Achieving stable, sustained fusion reactions would provide a nearly limitless supply of clean energy. Advanced extraterrestrial civilizations may have already mastered fusion technology, using it to propel their spacecraft across interstellar and intergalactic distances.

Ancient Alien Theory proponents argue that evidence of such advanced energy production techniques can be found in ancient texts and artifacts, suggesting that these extraterrestrial visitors shared their knowledge of fusion power with ancient civilizations. This could explain the existence of advanced technologies in the ancient world that appear to be far beyond the capabilities of the known cultures of the time.

Dark matter and dark energy are two mysterious components of the universe, comprising approximately 95% of its total mass-energy content. While their true nature remains unknown, they represent another potential source of exotic energy for advanced propulsion systems. If an extraterrestrial civilization could harness the power of dark matter and dark energy, it could potentially achieve intergalactic travel with ease.

Advocates of Ancient Alien Theory contend that ancient accounts of otherworldly beings and their advanced technologies may be attributed to these civilizations' mastery of dark matter and dark energy propulsion. By unlocking the secrets of these elusive cosmic phenomena, they could have traversed the universe, visiting Earth and leaving their mark on human history.

The potential for antimatter propulsion and other exotic energy sources to enable intergalactic travel lends further credence to the Ancient Alien Theory. Advanced extraterrestrial civilizations may have mastered these energy sources, allowing them to journey to Earth and share their knowledge with ancient peoples. As our understanding of these energy sources continues to grow, we may one day follow in their footsteps and explore the vast reaches of the cosmos.

In examining ancient artifacts and structures, there is often evidence of advanced technology and engineering that seems beyond the capabilities of human civilizations at the time. Ancient Alien Theory proponents suggest that artificial intelligence and robotics may have played a role in these achievements, as extraterrestrial visitors could have provided technological assistance to our ancestors.

For example, the precise construction of the Egyptian pyramids and the intricate carvings of Puma Punku in Bolivia have long puzzled researchers. Ancient Alien Theory proponents argue that these feats could be attributed to the use of advanced robotics and AI-driven machines provided by extraterrestrial visitors.

Many ancient texts and legends describe encounters with non-human entities possessing extraordinary abilities and knowledge. These accounts often depict beings with superior intellect and technological prowess, which could be interpreted as evidence of AI-driven or robotic visitors from other worlds.

For instance, the ancient Indian epic, the Mahabharata, tells of powerful weapons and flying machines called vimanas. These accounts could be interpreted as descriptions of advanced AI-driven spacecraft or robotic emissaries from extraterrestrial civilizations. Similarly, the Greek myth of Hephaestus, the god of fire and metalworking, describes a skilled craftsman who created automatons to aid him in his work. This could be seen as an early account of an encounter with robotic or AI-driven technology provided by ancient alien visitors.

The influence of artificial intelligence and robotics in our modern world is undeniable. From manufacturing and medicine to space exploration and even art, AI-driven systems have become an integral part of our lives. Ancient Alien Theory proponents argue that this rapid advancement in technology could be attributed, at least in part, to the influence of extraterrestrial civilizations that visited Earth in the past.

By providing ancient civilizations with advanced knowledge and technology, these visitors may have laid the foundation for our current understanding of AI and robotics. This idea is supported by the numerous ancient accounts of otherworldly beings imparting wisdom and technological expertise to humans, setting the stage for the technological revolution we are experiencing today.

As we continue to develop increasingly sophisticated AI-driven systems and robotics, our ability to explore and understand the universe will expand exponentially. These advancements will enable us to undertake longer and more complex missions, potentially leading to the discovery of extraterrestrial life and further supporting the Ancient Alien Theory.

Furthermore, as we continue to explore the cosmos, we may uncover additional evidence of ancient alien visitations in the form of AI-driven probes or robotic artifacts left behind by extraterrestrial civilizations. This ongoing search for extraterrestrial intelligence and the role of AI and robotics in intergalactic travel will undoubtedly play a crucial role in our understanding of ancient aliens and their impact on human history.

The Fermi Paradox and the search for extraterrestrial intelligence have provided fertile ground for proponents of the Ancient Alien Theory. They argue that the paradox does not preclude the existence of advanced extraterrestrial civilizations, but rather suggests that humanity may not have recognized or fully understood the evidence of their existence.

Supporters of the Ancient Alien Theory contend that evidence of extraterrestrial civilizations can be found in the remnants of ancient cultures, including artifacts, structures, and texts. They propose that these remnants demonstrate the presence of advanced alien technologies and knowledge that could not have been developed by humans at that time. This evidence, they argue, offers a more plausible explanation for the Fermi Paradox, as it indicates that extraterrestrial civilizations have already made contact with Earth and influenced human history.

Another argument put forth by proponents of the Ancient Alien Theory is that advanced alien civilizations may have deliberately chosen to avoid overt contact with humanity, either due to a policy of non-interference or for other reasons. This stealthy approach to contact could help explain the Fermi Paradox, as extraterrestrial civilizations might be actively observing and influencing human affairs from a distance without revealing their presence.

Ancient Alien Theory supporters also argue that advanced alien civilizations may have developed methods of intergalactic travel that remain beyond our current understanding of physics and technology. These advanced methods of travel could enable extraterrestrial civilizations to traverse vast interstellar distances, allowing them to visit Earth and influence human history. The existence of such technologies would provide a possible resolution to the Fermi Paradox, as it would suggest that advanced civilizations are indeed present in the universe, but have simply not been detected due to their ability to travel and interact with other worlds in ways that remain undetectable by our current technological capabilities.

In light of the arguments presented by proponents of the Ancient Alien Theory, it becomes apparent that the Fermi Paradox should not be viewed as an insurmountable obstacle to the existence of advanced extraterrestrial civilizations. Instead, the paradox may serve as a catalyst for rethinking our approach to the search for extraterrestrial intelligence and reconsidering the evidence of ancient alien contact. By embracing the possibility that advanced civilizations have already visited Earth and influenced human history, we may gain new insights into the nature of the universe and our place within it.

The Drake Equation takes into account several factors, including the rate of star formation in our galaxy, the fraction of stars with planetary systems, the number of planets capable of supporting life, the fraction of those planets that develop intelligent life, the fraction of intelligent civilizations that develop detectable communication technology, and the lifetime of such civilizations. By considering these variables, the Drake Equation provides a framework for estimating the potential number of extraterrestrial civilizations capable of interstellar communication.

Proponents of the Ancient Alien Theory argue that the potential existence of numerous extraterrestrial civilizations throughout the cosmos, as suggested by the Drake Equation, lends credence to the idea that Earth may have been visited by ancient aliens in the past. They contend that these advanced beings could have influenced the development of human civilization by sharing their knowledge and technology.

The presence of otherworldly beings and advanced technology in ancient myths and legends can be seen as supporting the Ancient Alien Theory. Many ancient cultures have stories describing encounters with god-like beings descending from the heavens, imparting wisdom and guiding the course of human history. These accounts, such as the Sumerian Anunnaki, the Egyptian gods, and the Mayan stories of Kukulkan, can be interpreted as evidence of extraterrestrial contact in our distant past.

Many ancient structures, such as the Egyptian pyramids and Stonehenge, display a high level of astronomical knowledge and precision, suggesting that the builders may have had access to advanced technology and knowledge. Ancient Alien Theory proponents argue that this advanced knowledge may have been imparted by extraterrestrial visitors, who used their understanding of astronomy and mathematics to help construct these monumental structures.

The potential existence of numerous extraterrestrial civilizations, as suggested by the Drake Equation, implies that the likelihood of Earth having been visited by ancient aliens is higher than previously thought. This idea is supported by the wealth of ancient myths, legends, and architectural evidence suggesting advanced knowledge and technology. As our understanding of the cosmos expands and we uncover more information about the potential for extraterrestrial life, the Ancient Alien Theory continues to gain traction as a plausible explanation for the mysteries of human history and the development of civilization.

As our understanding of the universe expands, so does the potential for discovering ways to circumvent the limitations imposed by the speed of light. Theoretical physics continues to explore concepts such as warp drives, tachyons, and other faster-than-light phenomena that could hold the key to traversing intergalactic distances. Proponents of the Ancient Alien Theory argue that advanced extraterrestrial beings may have already mastered these concepts, leaving behind evidence of their visitations in the form of ancient structures, artifacts, and knowledge.

The exploration of wormholes as a viable method for intergalactic travel remains a crucial area of research for scientists. By stabilizing and manipulating these theoretical structures, we may be able to create shortcuts through spacetime, enabling us to explore the far reaches of the universe. Ancient Alien Theory supporters suggest that extraterrestrial civilizations could have used wormholes to visit Earth, inspiring ancient myths and legends of otherworldly beings and bestowing advanced knowledge upon our ancestors.

Quantum physics, with its inherently strange and counterintuitive properties, may hold the key to unlocking new methods of intergalactic travel. Concepts such as quantum entanglement and quantum teleportation, while currently limited in scope, have the potential to revolutionize our understanding of space and time. Ancient Alien Theory proponents argue that advanced extraterrestrial civilizations may have harnessed the power of quantum physics to visit Earth, providing ancient societies with knowledge and technologies far beyond their time.

As our ability to search for and communicate with extraterrestrial civilizations improves, the likelihood of discovering advanced beings capable of intergalactic travel increases. Projects like the Breakthrough Listen initiative and the development of new, powerful telescopes offer the potential to identify and study distant planets that could harbor intelligent life. The Ancient Alien Theory suggests that we may find evidence of past visitations and interactions with these civilizations in the form of ancient texts, artifacts, and architectural marvels.

The study of potential ancient alien encounters may inspire the development of advanced technologies that could enable intergalactic travel. By examining the evidence left behind by these hypothetical extraterrestrial visitors, researchers may be able to reverse-engineer their technology, paving the way for humanity's expansion into the cosmos. The Ancient Alien Theory not only challenges our understanding of human history but also provides a potential roadmap for the future of intergalactic travel and our place in the universe.

As humanity moves closer to realizing the dream of intergalactic travel, the Ancient Alien Theory serves as a reminder of the potential for collaboration and shared knowledge between civilizations. By examining the evidence of ancient alien visitations and embracing the possibility that advanced extraterrestrial beings have influenced our past, we can pave the way for a future where intergalactic exploration and cooperation become a reality.

In conclusion, the exploration of intergalactic travel theories and their potential implications for the existence of ancient aliens offers a captivating look into the possibilities of human space exploration and our place in the cosmos. As we examine the various methods for traversing the immense distances between galaxies, we can better appreciate the potential existence of ancient alien civilizations and their influences on the development of human history. By pushing the boundaries of our understanding of physics, engineering, and technology, we may one day find conclusive evidence of ancient alien visitations, fundamentally transforming our perception of our position in the universe.

The scientific pursuit of intergalactic travel not only exemplifies human curiosity and innovation but also serves as a vehicle for exploring the possibility of ancient alien civilizations and their impact on our past. The Ancient Alien Theory, with its thought-provoking and influential ideas, will continue to drive our quest for answers about our origins and the enigmas of the cosmos.

As we delve deeper into the mysteries of space, developing new propulsion technologies and expanding our knowledge of the universe, the likelihood of intergalactic travel and the existence of ancient aliens becomes increasingly plausible. By taking into account various theories and technological advancements that could facilitate intergalactic travel, we open ourselves up to the possibility of following in the footsteps of our ancient alien visitors, journeying to the stars and uncovering untold wonders of the universe.

Unraveling the connection between intergalactic travel and ancient alien theories can bring us closer to understanding the broader implications of our search for extraterrestrial life. As we develop new ways to reach distant galaxies and expand our knowledge of the cosmos, we may find that the wisdom and experiences of ancient civilizations can provide invaluable insights into our own future among the stars.

In this exciting era of discovery, we stand on the precipice of incredible advancements in our understanding of the universe and our capabilities for intergalactic travel. The potential to uncover evidence of ancient alien visitations and to validate the theories that have captivated our imaginations for generations is now within our grasp.

As we continue our pursuit of knowledge, bridging the gap between science and the Ancient Alien Theory, we have the opportunity to redefine our understanding of human history and our place in the cosmos. By embracing the possibilities presented by intergalactic travel and ancient alien civilizations, we may one day unlock the secrets of the universe, forging a new path toward a future filled with untold wonders and discoveries.

Chapter 18

Extraterrestrial Spacecraft

Designs and Technology

The possibility of extraterrestrial life has captivated human imagination for centuries, and the concept of advanced alien civilizations visiting Earth in technologically sophisticated spacecraft remains a central theme within Ancient Alien Theory. In this chapter, we will explore various aspects of extraterrestrial spacecraft, including their designs, technology, and potential interactions with ancient human civilizations. To evaluate the plausibility of such claims, we will consider scientifically accurate information, documented sightings, and recovered artifacts that suggest extraterrestrial involvement in our past.

Ancient texts from various cultures around the world contain numerous accounts of celestial phenomena and otherworldly beings that proponents of Ancient Alien Theory argue are indicative of extraterrestrial spacecraft visitations. These texts often describe aerial vehicles, advanced technology, and encounters with beings that appear to possess knowledge and capabilities far beyond those of contemporary human civilizations.

For instance, the Sumerian texts from Mesopotamia, considered some of the oldest written records in human history, contain stories of the Anunnaki, a race of gods said to have descended from the heavens. These beings reportedly possessed advanced knowledge and technology, which they shared with ancient humans. Similarly, the Bible's Book of Ezekiel contains a famous passage describing a vision of a "wheel within a wheel" and beings descending from the sky, which some interpret as a depiction of an extraterrestrial spacecraft.

The examination of ancient artwork and iconography has revealed numerous examples of objects and beings that resemble modern UFO sightings and descriptions of extraterrestrial visitors. In many cases, these depictions appear in religious or spiritual contexts, suggesting that ancient cultures may have interpreted these encounters as divine or supernatural events.

For example, ancient Egyptian artwork often features disc-shaped objects hovering in the sky, sometimes accompanied by beings with elongated skulls and large eyes that resemble the stereotypical "grey alien" image. In pre-Columbian Mesoamerican cultures, such as the Maya and Aztec, there are numerous representations of what appear to be humanoid beings wearing suits and helmets, which some believe to be depictions of extraterrestrial visitors. These examples suggest that encounters with otherworldly beings and their advanced technology may have been a common theme in ancient cultures.

Many ancient megalithic structures and monuments around the world display advanced construction techniques, precise astronomical alignments, and other features that proponents of Ancient Alien Theory argue are indicative of extraterrestrial involvement. They contend that the knowledge and technology required to build these structures were beyond the capabilities of the civilizations that created them, and therefore must have been provided by advanced extraterrestrial visitors.

For example, the Great Pyramid of Giza in Egypt demonstrates remarkable precision in its construction, with its sides aligned to true north within a fraction of a degree. Similarly, Stonehenge in England and other megalithic sites around the world display alignments with specific celestial events, such as solstices and equinoxes, suggesting advanced astronomical knowledge.

Countless legends and folklore from diverse cultures around the world recount encounters with otherworldly beings and their advanced technology. These stories often describe gods or celestial beings descending from the sky in chariots, on winged creatures, or in other types of flying vehicles.

In Greek mythology, for example, the gods were said to reside on Mount Olympus and travel through the sky in chariots. The Nordic legends of the gods known as the Aesir describe their home in the realm of Asgard, connected to the human world through the Bifrost, a celestial bridge. Chinese mythology is replete with tales of flying dragons that carried celestial beings from the heavens.

These widespread legends and folklore provide further evidence that ancient human civilizations may have had encounters with extraterrestrial spacecraft and their occupants, interpreting these events through the lens of their cultural and religious beliefs.

The Vimanas are an intriguing aspect of ancient Indian mythology, as they represent a sophisticated understanding of flight and aerodynamics in a time when such concepts were thought to be unknown. Described in the Vedas, as well as in other Sanskrit texts such as the Mahabharata and the Ramayana, these flying machines are said to have been used by the gods and other celestial beings to traverse the heavens and visit the Earth.

The ancient texts provide detailed accounts of various types of Vimanas, with descriptions suggesting the existence of diverse designs and capabilities. Some Vimanas were said to resemble simple chariots or aerial vehicles, while others were described as advanced aircraft with the ability to travel vast distances, not only within Earth's atmosphere but also across interstellar space. This wide range of capabilities implies that the creators of these Vimanas possessed a deep understanding of aeronautical engineering and related technologies.

The propulsion systems of Vimanas, as described in the ancient texts, were quite advanced for their time. One such propulsion system is the

"Mercury Vortex Engine," which involves the use of mercury as a fuel source. The engine purportedly harnessed the power of mercury to create a vortex, generating lift and allowing the Vimana to fly at high speeds. Although this technology remains unproven in modern scientific research, the concept of a mercury-powered engine demonstrates the advanced technological understanding of those who wrote about Vimanas in ancient times.

The existence of Vimanas in ancient Indian mythology suggests the possibility that advanced extraterrestrial beings may have visited Earth in the past, sharing their knowledge of flight and advanced technology with the people they encountered. The detailed descriptions of Vimanas and their capabilities, along with the understanding of advanced propulsion systems, support the idea that these flying machines could have been inspired by extraterrestrial technology.

The descriptions of Vimanas in ancient Indian texts are consistent with the broader narrative of Ancient Alien Theory, which proposes that extraterrestrial beings visited Earth and influenced the development of human civilizations. Vimanas could represent evidence of this ancient contact, as their advanced designs and propulsion systems are difficult to explain within the context of known human technology during that era. By considering the existence of Vimanas as potential evidence of ancient alien encounters, we can further explore the possibility that advanced extraterrestrial beings played a significant role in shaping human history and technological development.

The prevalence of artwork from ancient cultures that seemingly depicts UFOs and extraterrestrial beings has long been a point of interest for proponents of Ancient Alien Theory. These visual records, found in various forms such as paintings, sculptures, and tapestries, suggest that the idea of extraterrestrial spacecraft and otherworldly visitors has been woven into the fabric of human culture for centuries.

The roots of ancient artwork featuring UFO-like objects can be traced back to prehistoric cave paintings. In various locations around the world, including sites in France, Australia, and Africa, cave paintings have been discovered that depict disc-shaped objects and humanoid figures wearing what appear to be helmets or suits. These images suggest that even our most ancient ancestors may have been aware of or encountered extraterrestrial visitors and their advanced technology.

In the ancient Egyptian civilization, numerous artifacts and hieroglyphs depict disc-shaped objects and beings with elongated skulls or other non-human features. One of the most notable examples is the Temple of Seti I in Abydos, where a series of hieroglyphs appears to show modern technological devices such as helicopters and submarines. While some scholars argue that these depictions are merely the result of overlapping hieroglyphs, others maintain that they provide evidence of extraterrestrial encounters in ancient Egypt.

During the Renaissance and Medieval periods, numerous paintings and tapestries featured objects in the sky that bear a striking resemblance to modern UFO sightings. For example, in the painting "The Madonna with Saint Giovannino" by Domenico Ghirlandaio, a disc-shaped object can be seen in the background, with a witness on the ground seemingly gazing at it. Similarly, the "Annunciation with Saint Emidius" by Carlo Crivelli shows a beam of light emanating from a circular object in the sky, shining down upon the Virgin Mary. These visual records from different time periods and cultures add to the growing body of evidence suggesting that extraterrestrial spacecraft have been present in human culture for centuries.

Mesoamerican cultures, such as the Maya and the Aztecs, have also produced artwork and architecture that seemingly depicts extraterrestrial encounters. The ancient Mayan site of Palenque, for example, features a relief carving that many argue shows the ruler Pakal seated in a spacecraft-like vehicle. Additionally, the Aztec god Quetzalcoatl is often portrayed as a feathered serpent, which some interpret as a symbolic representation of an extraterrestrial spacecraft.

By examining these visual records from ancient cultures around the world, it becomes increasingly apparent that the idea of extraterrestrial spacecraft and otherworldly visitors has been a consistent theme throughout human history. The existence of such artwork across disparate civilizations suggests that these depictions may be more than mere coincidence or artistic license; rather, they may provide compelling evidence for the existence of ancient aliens and their interactions with humanity.

The concept of antigravity technology has captured the imagination of both scientists and Ancient Alien Theory proponents. This theoretical propulsion system is based on the manipulation of gravitational forces to create movement without the need for conventional rocket engines. By generating an opposing gravitational field, a spacecraft could potentially lift off and travel through space with remarkable ease. The behavior of purported UFOs, such as their rapid acceleration, ability to hover, and sudden changes in direction, lends credibility to the idea that antigravity technology could be in use by extraterrestrial spacecraft.

The notion of faster-than-light travel is considered essential for any civilization seeking to engage in interstellar travel within reasonable timeframes. This idea has led to the development of warp drive theories, which propose that a spacecraft could move faster than the speed of light by warping the fabric of spacetime. Proponents of Ancient Alien Theory argue that extraterrestrial spacecraft visiting Earth would likely possess this advanced capability. The concept of the Alcubierre drive, a theoretical framework for warp drive technology proposed by Mexican physicist Miguel Alcubierre, offers a scientific basis for such a propulsion system. Although the Alcubierre drive remains purely theoretical at this time, its potential applications are consistent with the capabilities attributed to extraterrestrial spacecraft.

Another possibility for advanced propulsion systems involves the use of exotic matter, a hypothetical form of matter with properties that defy conventional physics. One example of exotic matter is negative mass, which, according to theoretical predictions, would exhibit repulsive gravitational effects. If extraterrestrial spacecraft were capable of harnessing and manipulating exotic matter, they could potentially generate propulsion through the interaction between negative and positive mass. This method could allow for efficient and rapid interstellar travel, aligning with the capabilities ascribed to alien spacecraft by Ancient Alien Theory proponents.

The field of breakthrough propulsion physics focuses on the development of innovative propulsion concepts that could revolutionize space travel. These concepts often involve the application of advanced or exotic physics principles, such as manipulating spacetime, harnessing the energy of the quantum vacuum, or creating artificial gravity fields. Ancient Alien Theory supporters posit that extraterrestrial civilizations may have already mastered these breakthrough propulsion technologies, enabling them to traverse the vast distances between stars with ease. The existence of such advanced propulsion systems would help to explain the remarkable capabilities exhibited by UFOs and other purported extraterrestrial spacecraft.

The exploration of advanced propulsion systems is an integral aspect of Ancient Alien Theory, as it provides a potential explanation for the presence of extraterrestrial spacecraft on Earth. By considering the possibility of antigravity technology, warp drives, exotic matter propulsion, and breakthrough propulsion physics, we can better understand the capabilities that alien spacecraft might possess. Although our current understanding of these propulsion systems is limited, their potential applications align with the remarkable characteristics attributed to extraterrestrial spacecraft, supporting the notion that ancient aliens may have indeed visited our planet.

The concept of antigravity technology as a form of advanced propulsion is a popular notion among Ancient Alien Theory proponents, who argue that such technology would explain the extraordinary capabilities of purported extraterrestrial spacecraft. Antigravity technology involves manipulating gravitational forces to counteract the effects of gravity, allowing a vehicle to lift off, hover, and travel through space without the need for conventional propulsion systems.

The idea that UFOs utilize antigravity technology is based on the observation of their unique flight characteristics, which seemingly defy the laws of physics. Witnesses have reported UFOs performing rapid accelerations, abrupt changes in direction, and hovering motionless in the sky, suggesting that these craft possess propulsion systems far beyond human technology. The use of antigravity technology would enable such maneuvers without the need for large, fuel-consuming engines or loud, disruptive sound emissions.

Ancient texts and legends from various cultures describe flying machines and objects that seemingly defy gravity. For example, the Indian Vimanas, as mentioned earlier, were described as being able to hover and move in any direction, much like modern UFO sightings. The existence of antigravity technology in ancient times would support the Ancient Alien Theory, which proposes that advanced extraterrestrial beings visited Earth and shared their knowledge with humans.

While antigravity technology remains a theoretical concept, some researchers have proposed scientific models to explain its potential functionality. One such concept is the manipulation of the Higgs field, a quantum field responsible for giving particles mass, which in turn generates gravitational forces. By altering the Higgs field, it may be possible to reduce or nullify the mass of an object, effectively counteracting the effects of gravity. If extraterrestrial civilizations have discovered a way to manipulate the Higgs field, their spacecraft would be capable of the seemingly impossible feats often attributed to UFOs.

Scientists and engineers have been investigating the possibility of antigravity technology for decades. Some experimental research has focused on the development of devices that generate lift through unconventional means, such as rotating superconductors or high-frequency electromagnetic fields. While these experiments have yet to produce definitive proof of antigravity technology, they demonstrate the continued interest in this field and the potential for future breakthroughs.

Antigravity technology remains a fascinating and controversial area of study within the context of Ancient Alien Theory. While current scientific understanding has not yet produced a working antigravity device, the potential implications of such technology are profound. If extraterrestrial beings possess antigravity technology, it would explain the remarkable flight characteristics of UFOs and support the notion of advanced alien civilizations visiting Earth in the past.

Faster-than-light (FTL) travel is considered a crucial requirement for interstellar travel, as it enables the traversal of vast cosmic distances within humanly comprehensible timeframes. The vastness of space means that even at the speed of light, traveling between star systems would take years, if not decades or centuries. Ancient Alien Theory proponents argue that extraterrestrial civilizations capable of visiting Earth must possess advanced FTL propulsion systems to make such journeys feasible.

The warp drive is a hypothetical propulsion system that would allow a spacecraft to travel faster than the speed of light by manipulating the fabric of spacetime itself. This concept, inspired by the general theory of relativity, postulates that instead of moving through space, a spacecraft could warp spacetime around itself, creating a "bubble" that allows it to cover vast distances without violating the cosmic speed limit. This method of propulsion would effectively shorten the space between two points, enabling the spacecraft to traverse these distances without needing to accelerate to relativistic speeds.

Mexican physicist Miguel Alcubierre proposed the Alcubierre drive as a theoretical framework for warp drive technology in 1994. The Alcubierre drive envisions a spacecraft surrounded by a ring-like structure that generates a spacetime distortion, creating a wave-like effect that propels the craft forward. This distortion causes spacetime to contract in front of the spacecraft and expand behind it, effectively allowing the craft to "surf" on the spacetime wave.

The concept of the Alcubierre drive has gained traction within the scientific community, and several researchers have explored its potential implications and feasibility. While the idea remains purely theoretical at this time, advancements in our understanding of spacetime, dark energy, and the manipulation of gravitational forces could eventually lead to the development of this technology.

The idea of warp drive technology and FTL travel aligns well with the capabilities attributed to extraterrestrial spacecraft by proponents of Ancient Alien Theory. The existence of advanced FTL propulsion systems would provide a means for alien civilizations to traverse the vast distances between stars and interact with ancient human societies. This technology could also explain the seemingly impossible maneuvers and rapid acceleration observed in some modern UFO sightings, which defy our current understanding of physics and aerospace engineering.

The development of warp drive technology or other forms of FTL propulsion would revolutionize human space exploration, allowing us to reach distant star systems and potentially interact with extraterrestrial civilizations. Studying the possibilities of FTL travel, inspired by Ancient Alien Theory and the alleged capabilities of extraterrestrial spacecraft, could lead to significant advancements in propulsion technology and open up new horizons for humanity's understanding of the cosmos.

The immense energy demands associated with interstellar travel necessitate the development and utilization of advanced power generation techniques that far surpass the potential of traditional fuel sources. Proponents of Ancient Alien Theory contend that extraterrestrial spacecraft are likely to employ innovative energy sources, including zero-point energy and nuclear fusion, to provide the energy needed for propulsion systems and other advanced technology.

Zero-point energy is a hypothetical energy source derived from the quantum vacuum, which is the lowest energy state of a quantum system. Tapping into zero-point energy could potentially offer an almost inexhaustible power supply, enabling spacecraft to traverse vast interstellar distances without relying on conventional fuels. The concept of zero-point energy remains a subject of ongoing debate within the scientific community. However, if extraterrestrial spacecraft were to harness this energy, their advanced capabilities would align with the descriptions provided by Ancient Alien Theory proponents. Additionally, the ability to utilize zero-point energy could have been shared with ancient civilizations, potentially explaining the construction of enigmatic structures that seemingly required advanced technology.

Nuclear fusion, a process involving the combination of atomic nuclei to release energy, has long been viewed as a potential power source for sophisticated spacecraft propulsion. Fusion reactions generate considerably more energy than conventional nuclear fission while producing minimal radioactive waste, rendering them an ideal candidate for extended space travel. If extraterrestrial civilizations have successfully harnessed nuclear fusion, their spacecraft could possess advanced propulsion systems and energy generation capabilities that dramatically surpass those of current human technology.

The efficient storage and distribution of energy are crucial aspects of advanced power generation systems. Extraterrestrial spacecraft might employ cutting-edge energy storage technologies, such as high-capacity supercapacitors or advanced battery designs, to ensure a reliable and consistent energy supply. These advanced energy storage systems would enable spacecraft to maintain functionality during long-duration interstellar journeys and facilitate the use of energy-intensive technologies. Moreover, ancient civilizations may have benefited from contact with these advanced energy storage and distribution systems, leading to the development of technologies such as the Baghdad Battery, which some researchers believe to be an early form of a galvanic cell.

Apart from zero-point energy and nuclear fusion, extraterrestrial spacecraft might exploit other cosmic energy sources to power their systems. For instance, these craft could utilize solar energy through highly efficient solar panels, capturing and converting a greater percentage of the sun's energy than current human technology allows. Additionally, extraterrestrial spacecraft might harness energy from other celestial bodies, such as stars or even black holes, using advanced energy conversion methods that remain beyond our current understanding.

By considering the potential advanced energy sources and power generation methods employed by extraterrestrial spacecraft, Ancient Alien Theory proponents suggest that these craft possess capabilities far beyond those of contemporary human technology. Furthermore, these advanced energy systems could have played a significant role in shaping the development of ancient civilizations, providing them with access to powerful energy sources that facilitated the construction of complex structures and the development of advanced technologies.

The concept of zero-point energy is rooted in the principles of quantum mechanics, which suggest that even in the lowest possible energy state, a quantum system still contains some residual energy. This energy, known as zero-point energy, is considered a theoretical and potentially limitless power source that could revolutionize spacecraft propulsion systems and other forms of technology.

The idea of extraterrestrial spacecraft utilizing zero-point energy as a power source aligns with the advanced capabilities often attributed to these craft by proponents of Ancient Alien Theory. By harnessing the immense energy of the quantum vacuum, extraterrestrial civilizations could achieve unprecedented levels of efficiency and sustainability in their spacecraft propulsion systems, enabling them to traverse vast interstellar distances without relying on conventional fuel sources.

In addition to enabling long-distance space travel, zero-point energy could also provide extraterrestrial spacecraft with a myriad of other technological advantages. For instance, the use of zero-point energy could lead to the development of advanced shielding systems that protect against radiation, meteoroids, and other hazards of deep-space travel. Furthermore, zero-point energy could potentially facilitate the creation of advanced life-support systems, allowing extraterrestrial beings to maintain a habitable environment within their spacecraft over extended periods.

The application of zero-point energy in extraterrestrial spacecraft technology also has implications for the broader Ancient Alien Theory. If these advanced civilizations were able to harness the power of the quantum vacuum, they could have shared this knowledge with ancient human civilizations, potentially leading to the development of advanced technologies that remain unexplained to this day. Ancient artifacts, such as the Baghdad Battery, which some researchers argue could be an early form of an electrical device, might represent early attempts by humans to apply the principles of zero-point energy in their own technology.

While zero-point energy remains a subject of debate and continued research within the scientific community, its potential applications in extraterrestrial spacecraft technology offer compelling support for the advanced capabilities attributed to these craft by Ancient Alien Theory proponents. As our understanding of quantum mechanics and the nature of the quantum vacuum continues to evolve, the possibility of harnessing zero-point energy as a power source for advanced spacecraft and other technologies may become increasingly plausible, further strengthening the case for the existence of ancient aliens and their interactions with human civilizations.

Nuclear fusion holds great promise as a source of energy for advanced spacecraft propulsion due to its ability to produce a tremendous amount of energy from a relatively small amount of fuel. This energy output far exceeds that of traditional nuclear fission and chemical combustion, making it an ideal candidate for long-duration space travel. Ancient Alien Theory proponents argue that extraterrestrial civilizations may have already harnessed the power of nuclear fusion, giving them the ability to traverse vast interstellar distances and visit Earth in the past.

Supporters of Ancient Alien Theory point to ancient texts and artifacts that suggest the use of advanced technology, including nuclear fusion, by extraterrestrial visitors. For instance, certain descriptions in ancient Indian texts, such as the Mahabharata, hint at the possibility of powerful weapons and energy sources that bear a resemblance to nuclear reactions. These accounts may be interpreted as evidence of extraterrestrial civilizations possessing advanced knowledge of nuclear fusion, which they could have used to power their spacecraft and other technology.

The mastery of nuclear fusion by extraterrestrial civilizations would have significant implications for the design and capabilities of their spacecraft. A fusion-powered spacecraft would be able to travel vast distances without the need for refueling, as the fuel required for fusion reactions is abundant in the universe (e.g., hydrogen isotopes like deuterium and tritium). The efficiency and power generated by nuclear fusion would enable extraterrestrial spacecraft to achieve extremely high speeds, potentially allowing them to travel between stars within a relatively short timeframe. This level of technological sophistication could explain the advanced capabilities attributed to extraterrestrial spacecraft by Ancient Alien Theory proponents.

The potential existence of nuclear fusion-powered extraterrestrial spacecraft raises intriguing questions about the impact of such technology on human civilization. If ancient aliens visited Earth with fusion-powered craft, they may have shared knowledge about this advanced energy source with ancient human societies. This knowledge transfer could have influenced the development of human technology, potentially leading to advancements in energy generation, propulsion, and other fields.

Moreover, the pursuit of nuclear fusion as a viable energy source for human use has been an ongoing scientific endeavor, with researchers working tirelessly to achieve sustained, controlled fusion reactions. The potential success of these efforts could revolutionize human space travel and energy production, bringing us closer to the advanced capabilities exhibited by extraterrestrial civilizations. The existence of nuclear fusion-powered spacecraft in ancient times, as suggested by Ancient Alien Theory, serves as a powerful reminder of the possibilities that await us as we continue to push the boundaries of scientific understanding and technological innovation.

The idea that extraterrestrial spacecraft are composed of advanced materials with unique properties is a central tenet of Ancient Alien Theory. The utilization of such materials in the construction of these craft would enable them to withstand the harsh conditions of space travel and provide enhanced performance characteristics. These materials might include exotic alloys or composites that exhibit exceptional strength, lightweight construction, or resistance to extreme temperatures and radiation.

One of the most intriguing potential properties of extraterrestrial spacecraft materials is the ability to self-heal. Materials with self-healing capabilities can autonomously repair damage, extending the lifespan and durability of the spacecraft. This advanced feature could be particularly beneficial for long-duration space missions or encounters with debris, meteoroids, or other hazards in space. By incorporating self-healing materials into their design, extraterrestrial spacecraft may demonstrate increased resilience and adaptability, surpassing the capabilities of current human-made vehicles.

Lightweight construction is another important aspect of extraterrestrial spacecraft materials. By utilizing advanced materials with high strength-to-weight ratios, these craft could achieve a significant reduction in mass, allowing for improved maneuverability and reduced energy consumption. This would be especially advantageous for interstellar travel, as reduced mass directly translates to increased efficiency in propulsion systems. Some possible materials that may be employed in lightweight construction include carbon nanotubes, graphene, or other advanced nanomaterials.

Space is an extremely hostile environment, characterized by temperature extremes, vacuum conditions, and exposure to high levels of radiation. Extraterrestrial spacecraft would need to be constructed using materials capable of withstanding these harsh conditions to ensure the survival of their occupants and the integrity of the craft's systems. Advanced materials, such as radiation-resistant polymers, superinsulating ceramics, or specialized metal alloys, could be employed in the construction of these spacecraft to provide enhanced protection against the hazards of space travel.

In addition to the use of advanced materials, extraterrestrial spacecraft may be constructed using cutting-edge manufacturing techniques that are beyond our current technological capabilities. These advanced methods could include molecular nanotechnology, which allows for the precise manipulation of matter at the atomic scale, or advanced 3D printing technologies that enable the creation of complex structures and components with unparalleled precision. By utilizing these advanced manufacturing techniques, extraterrestrial civilizations may be able to construct spacecraft with extraordinary capabilities and performance characteristics.

The potential existence of extraterrestrial spacecraft constructed with advanced materials and techniques has significant implications for Ancient Alien Theory. Proponents argue that ancient human civilizations may have been influenced by these otherworldly visitors and their advanced technology. The presence of advanced materials in ancient artifacts, such as the purported "impossible" stone carving found in ancient Egypt, or the precise construction techniques observed in megalithic structures like Puma Punku, could be evidence of extraterrestrial involvement in our past. These interactions may have shaped the course of human history, introducing advanced knowledge and technology to ancient civilizations, and inspiring the myths and legends that have persisted throughout the ages.

Extraterrestrial spacecraft constructed using advanced materials would not only exhibit superior performance characteristics but also offer insights into the technological prowess of the civilizations that created them. The utilization of advanced materials, such as metamaterials and carbon nanotubes, would align with Ancient Alien Theory by supporting the notion that extraterrestrial beings possessed knowledge and technology far beyond that of ancient human civilizations.

Metamaterials, which are engineered materials with properties not typically found in nature, offer a wide range of potential applications in spacecraft design. For example, their unique electromagnetic properties could be employed to create advanced cloaking devices, rendering the spacecraft invisible to radar or even the naked eye. Such capabilities would explain the elusive nature of many purported UFO sightings and support the notion of extraterrestrial visitation.

Carbon nanotubes, another example of advanced materials, are known for their exceptional strength, flexibility, and lightweight construction. When integrated into the structural components of extraterrestrial spacecraft, carbon nanotubes would allow for increased durability and reduced weight, enabling these vehicles to withstand the extreme forces and stresses associated with interstellar travel. The use of carbon nanotubes in spacecraft design would also suggest that extraterrestrial beings possessed a deep understanding of nanotechnology, a field that has only recently been explored by human researchers.

In addition to metamaterials and carbon nanotubes, other advanced materials, such as graphene and exotic alloys with unique properties, could be employed in the construction of extraterrestrial spacecraft. These materials would provide further advantages, including enhanced thermal and electrical conductivity, radiation shielding, and resistance to extreme temperatures. The incorporation of such advanced materials into extraterrestrial spacecraft would support the idea that alien civilizations possessed technology far beyond the capabilities of ancient human societies, offering compelling evidence in favor of Ancient Alien Theory.

Self-healing materials possess unique properties that enable them to automatically repair damage caused by wear and tear, impacts, or exposure to harsh environments. These materials can be composed of various elements, including polymers, metals, and ceramics, and may feature embedded microcapsules, hollow fibers, or vascular networks that contain healing agents. When damage occurs, the healing agents are released and react with the damaged material, triggering a repair process that restores the material's structural integrity. This ability to self-repair can significantly extend the lifespan of a material and reduce the need for maintenance, making it an ideal choice for the construction of advanced extraterrestrial spacecraft.

Extraterrestrial spacecraft subjected to the rigors of long-duration space travel and exposure to extreme conditions, such as radiation, micrometeoroid impacts, and temperature fluctuations, would benefit greatly from the use of self-healing materials. These materials could be employed in various aspects of spacecraft design, including the outer hull, structural components, and internal systems, providing enhanced durability and resilience.

The use of self-healing materials in extraterrestrial spacecraft could also offer several advantages in terms of weight reduction and resource efficiency. By eliminating the need for redundant systems or heavy protective shielding, these advanced materials could enable spacecraft to be lighter and more fuel-efficient, thereby improving their overall performance and maneuverability.

The potential existence of self-healing materials in extraterrestrial spacecraft could provide further support for the Ancient Alien Theory by demonstrating the advanced technological capabilities of these craft. If ancient human civilizations had encountered extraterrestrial beings with such sophisticated materials, it could have influenced the development of their own construction techniques and inspired the creation of long-lasting structures that still stand today.

Furthermore, the discovery of artifacts or remnants of ancient structures composed of self-healing materials or displaying similar properties could be considered evidence of extraterrestrial influence in our past. As researchers continue to develop and explore the potential applications of self-healing materials, the potential connections between these advanced materials and the Ancient Alien Theory may become clearer, further challenging our understanding of human history and the potential existence of advanced extraterrestrial civilizations.

The potential existence of extraterrestrial spacecraft and their advanced communication and navigation systems has long been a topic of interest within Ancient Alien Theory. Proponents argue that the capabilities of these systems would far surpass those of human technology, enabling extraterrestrial beings to travel vast distances and interact with various celestial bodies, including Earth. In this section, we will explore several aspects of communication and navigation systems that may be employed by extraterrestrial spacecraft, focusing on concepts such as quantum communication, advanced artificial intelligence, and celestial navigation.

Quantum communication is a cutting-edge method of transmitting information that utilizes the principles of quantum mechanics. This approach offers numerous advantages over traditional communication methods, including increased security, resistance to interference, and the potential for instantaneous data transfer across vast distances. The use of quantum communication by extraterrestrial spacecraft would enable them to maintain contact with their home planets or other craft, facilitating efficient communication and coordination throughout the cosmos. Such advanced communication capabilities would support the idea that ancient aliens possessed knowledge and technology far beyond human understanding.

Advanced artificial intelligence (AI) is another crucial aspect of the communication and navigation systems that may be employed by extraterrestrial spacecraft. By utilizing AI systems with self-learning and adaptive capabilities, these craft could navigate and operate in complex environments without the need for direct control from their home planets. The presence of advanced AI in extraterrestrial spacecraft could also imply that ancient aliens possessed the ability to create and control intelligent machines, further supporting the notion that these beings were far more technologically advanced than humans.

Celestial navigation, the practice of using the position of celestial bodies to determine one's location and course, has been employed by human civilizations for millennia. However, the precision and accuracy required for interstellar travel would necessitate far more advanced celestial navigation techniques than those used by humans. Extraterrestrial spacecraft may employ sophisticated star-tracking systems or make use of gravitational waves to navigate the vast expanse of space. The potential use of advanced celestial navigation techniques by ancient aliens would further emphasize their superior technological capabilities and knowledge of the cosmos.

Another aspect of advanced navigation that may be employed by extraterrestrial spacecraft is multidimensional navigation, which involves the ability to travel through higher dimensions or alternate realities. While this concept may seem like science fiction, some theoretical physics models suggest that our universe may contain additional dimensions beyond the familiar three-dimensional space. If extraterrestrial beings have developed the ability to navigate through these higher dimensions, they would possess a level of technological prowess that far exceeds human understanding. The potential existence of multidimensional navigation capabilities among ancient aliens would provide additional support for the idea that these beings were capable of traversing vast distances and interacting with Earth and its inhabitants.

The advanced communication and navigation systems that may be employed by extraterrestrial spacecraft provide compelling evidence in favor of the existence of ancient aliens. The use of quantum communication, artificial intelligence, celestial navigation, and potentially even multidimensional navigation would demonstrate a level of technological sophistication that far surpasses human capabilities. As we continue to explore the depths of space and unlock the mysteries of our universe, the study of advanced communication and navigation systems within the context of Ancient Alien Theory will remain an important and thought-provoking area of inquiry.

Quantum communication, as a method of transmitting information using the principles of quantum mechanics, presents a fascinating possibility in the context of Ancient Alien Theory. The proponents of this theory argue that if extraterrestrial civilizations possess advanced technology, it is likely that their communication systems would also be highly sophisticated. Quantum communication would be an ideal candidate for such advanced systems, offering numerous benefits over traditional communication methods.

The main advantage of quantum communication lies in its ability to transfer information instantaneously across vast distances, a feature known as quantum entanglement. This phenomenon allows for particles to become entangled, sharing a unique connection that enables the instantaneous transfer of information, regardless of the distance between them. In the context of extraterrestrial spacecraft, this would allow for real-time communication between crafts and their home planets, even across the vast expanses of space. This would be a significant advantage over traditional communication methods, which are constrained by the speed of light and can take years to travel between stars.

Another benefit of quantum communication is its unparalleled security. Quantum information cannot be intercepted or tampered with without detection, as the act of observing or measuring the quantum state of a particle alters its properties. This inherent security would be highly advantageous for extraterrestrial civilizations, ensuring the privacy and integrity of their communications. Moreover, this secure method of communication could have potentially been shared with ancient human civilizations, offering advanced encryption techniques for the transmission of sensitive information.

Furthermore, the use of quantum communication by extraterrestrial spacecraft could provide insights into the advanced technological capabilities of these civilizations. If we were to intercept or detect traces of such communication, it could offer valuable information about the nature and intentions of these extraterrestrial visitors. In addition, the study of quantum communication in the context of Ancient Alien Theory could inspire further research into this cutting-edge field, potentially leading to the development of advanced communication systems for our own use.

The possibility that extraterrestrial spacecraft utilize quantum communication offers compelling support for the existence of ancient aliens. The advanced nature of this technology aligns with the high level of sophistication attributed to these civilizations by proponents of Ancient Alien Theory. Moreover, the potential benefits of quantum communication, such as instantaneous data transfer and unparalleled security, make it an ideal candidate for use by extraterrestrial civilizations in their interactions with each other and potentially with ancient human societies.

Extraterrestrial spacecraft equipped with advanced artificial intelligence (AI) systems would exhibit capabilities that far surpass those of contemporary human technology. These AI systems could be essential to the efficient and effective operation of the spacecraft, enabling them to traverse vast distances in space, adapt to new environments, and communicate with their home planets or other spacecraft.

The advanced AI systems on board extraterrestrial spacecraft would be responsible for navigating and exploring new environments, such as distant star systems or planetary surfaces. These AI systems could analyze and process vast amounts of data, such as the positions of celestial bodies, gravitational forces, and other physical properties of space, to determine the optimal course for their spacecraft. Furthermore, the AI systems could continuously adapt to new information, allowing the spacecraft to react to unexpected events or hazards, such as asteroid fields or extreme space weather, and to autonomously explore unknown environments.

Advanced AI systems aboard extraterrestrial spacecraft would be capable of making complex decisions and solving problems that might arise during their missions. These AI systems could evaluate numerous potential outcomes and select the best course of action, optimizing the spacecraft's performance and ensuring the success of their missions. For example, AI systems could be responsible for managing the spacecraft's energy consumption, balancing the use of various propulsion systems, and maintaining the integrity of the spacecraft's structure and systems in the face of potential damage or malfunctions.

Communication with extraterrestrial civilizations would likely be a critical aspect of the AI systems on board these spacecraft. Advanced AI systems could be responsible for encoding and decoding complex messages, facilitating communication between the spacecraft and its home planet or other craft, and deciphering information from other intelligent civilizations encountered during their missions. Additionally, the AI systems could help establish communication with ancient human civilizations, translating and interpreting the languages and knowledge systems of these cultures to facilitate the exchange of information.

In line with Ancient Alien Theory, the advanced AI systems on board extraterrestrial spacecraft could have played a pivotal role in the interactions between these beings and ancient human civilizations. These AI systems could have been responsible for analyzing and interpreting the cultural, linguistic, and technological characteristics of ancient human societies, allowing the extraterrestrial visitors to effectively communicate and collaborate with them. The AI systems could have also assisted in the transfer of knowledge and technology, teaching ancient civilizations about advanced mathematical concepts, astronomical phenomena, and engineering techniques that would have been instrumental in the development of their societies.

Extraterrestrial spacecraft visiting Earth in the past may have had a profound impact on ancient human civilizations, not only through direct technological assistance and knowledge transfer but also by influencing their cultural beliefs and practices. Proponents of Ancient Alien Theory suggest that various ancient mythologies and religious texts contain references to otherworldly beings, which could be interpreted as extraterrestrial visitors. These beings were often depicted as gods or divine entities, suggesting that their advanced technology and capabilities were interpreted as supernatural powers by ancient cultures.

The existence of architectural marvels from ancient civilizations has long been a cornerstone of the Ancient Alien Theory. Proponents argue that some of these structures, such as the Egyptian pyramids, Stonehenge, and the statues of Easter Island, display a level of precision and sophistication that would have been difficult to achieve without advanced knowledge or technology. It is suggested that extraterrestrial spacecraft and their occupants may have assisted in the construction of these structures, providing advanced tools, techniques, or direct labor to complete these seemingly impossible feats of engineering.

Some Ancient Alien Theory proponents speculate that extraterrestrial spacecraft and their occupants may have played a role in the development of human biology itself. This idea encompasses various theories, such as the notion that extraterrestrial beings may have genetically engineered or modified early humans, resulting in the rapid development of intelligence and physical traits that distinguish us from other primates. This concept of biological intervention is supported by the existence of ancient myths that describe gods or divine beings as creators of humanity or as playing a role in our evolution.

Artistic representations from ancient cultures worldwide provide intriguing evidence for the potential interaction between extraterrestrial spacecraft and ancient human civilizations. Many of these artworks, such as cave paintings, sculptures, and frescoes, appear to depict beings with non-human features, such as elongated skulls or large, almond-shaped eyes. In addition, some artistic depictions show what appears to be advanced technology, such as flying vehicles or advanced weaponry. Proponents of Ancient Alien Theory argue that these artistic representations provide evidence of extraterrestrial contact with ancient civilizations and that the artists were documenting their encounters with otherworldly beings and their technology.

The idea that extraterrestrial spacecraft may have visited Earth in the past and shared their advanced knowledge with ancient human civilizations raises the question of how this information was preserved and disseminated throughout history. Proponents of Ancient Alien Theory suggest that secret societies, religious institutions, and esoteric traditions may have played a role in safeguarding and transmitting this knowledge across generations. These organizations may have been established by the extraterrestrial visitors themselves or by ancient humans who were entrusted with their teachings. The preservation and dissemination of this knowledge could have contributed to the development of advanced technologies, scientific principles, and philosophical ideas that have shaped human history and culture.

The potential interaction between extraterrestrial spacecraft and ancient human civilizations leads to the possibility that such contact may have continued throughout history and into the present day. Proponents of Ancient Alien Theory argue that modern UFO sightings, encounters, and abduction experiences may be evidence of ongoing extraterrestrial contact with humanity. This ongoing relationship, they suggest, could be part of a larger plan or purpose that has yet to be fully understood or revealed.

The concept of knowledge transfer between extraterrestrial beings and ancient human civilizations is a cornerstone of Ancient Alien Theory. Proponents argue that these advanced beings shared their expertise in various fields with early humans, enabling them to create remarkable structures and develop sophisticated technologies that would have otherwise been beyond their capabilities.

One area where ancient aliens may have imparted their knowledge is in the fields of mathematics and astronomy. Evidence of this can be seen in the remarkable accuracy and complexity of ancient astronomical observatories, such as Stonehenge and the Mayan observatory at Chichen Itza. These structures demonstrate precise alignments with celestial bodies and a deep understanding of astronomical phenomena, such as solstices and eclipses. Additionally, the Sumerian civilization developed advanced mathematical concepts, including a base-60 numeral system that laid the foundation for modern timekeeping and geometry. Proponents of Ancient Alien Theory suggest that this advanced understanding of mathematics and astronomy may have been the result of extraterrestrial knowledge transfer.

Extraterrestrial beings may have also shared their advanced engineering and construction techniques with ancient human civilizations, enabling them to build complex structures that have withstood the test of time. The Great Pyramid of Giza, for example, is an engineering marvel that demonstrates precise measurements and an understanding of load-bearing principles that were far ahead of their time. Similarly, the megalithic site of Puma Punku in Bolivia features intricately carved stones with precise angles and interlocking features that defy conventional explanations of ancient construction methods. Ancient Alien Theory proponents argue that these advanced techniques could have been taught to early humans by extraterrestrial visitors.

Another area where ancient aliens may have shared their knowledge is in agriculture and medicine. The development of advanced agricultural techniques, such as crop rotation and irrigation, enabled ancient civilizations to support large populations and build thriving cities. Similarly, ancient cultures possessed a surprising degree of medical knowledge, including surgical procedures and the use of medicinal plants. The discovery of the ancient Sumerian medical text, known as the "Surgical Instructions of Nippur," suggests that early humans had a sophisticated understanding of anatomy and surgical techniques. Proponents of Ancient Alien Theory argue that these advancements in agriculture and medicine could have been the result of knowledge shared by extraterrestrial beings.

The development of language and writing systems is another aspect of ancient human civilization that could have been influenced by extraterrestrial beings. Many ancient cultures developed complex writing systems, such as cuneiform and hieroglyphics, that allowed them to record history, conduct trade, and share knowledge across generations. The similarities between some of these writing systems, despite being geographically distant and culturally distinct, suggest that they may have been influenced by a common source. Ancient Alien Theory proponents argue that extraterrestrial beings may have played a role in the development of these writing systems, providing early humans with the tools necessary for advanced communication and record-keeping.

The concept of knowledge transfer between extraterrestrial beings and ancient human civilizations is a key element of Ancient Alien Theory. This transfer of knowledge in areas such as mathematics, astronomy, engineering, agriculture, medicine, and language could have played a crucial role in the development of complex structures and advanced technologies in ancient civilizations. While the evidence supporting this theory is primarily circumstantial, it provides a thought-provoking explanation for the remarkable achievements of early human societies.

The idea that extraterrestrial beings provided direct technological assistance to ancient human civilizations is a captivating aspect of Ancient Alien Theory. It suggests that the seemingly impossible engineering feats and advanced artifacts of the past were not solely the product of human ingenuity but were achieved through collaboration with otherworldly visitors possessing superior knowledge and technology.

The construction of monumental structures, such as the Great Pyramid of Giza, Stonehenge, and the Moai statues on Easter Island, has long been a subject of wonder and debate. These structures display impressive engineering skills and a deep understanding of mathematics, astronomy, and geology that seem to surpass the capabilities of the civilizations responsible for their creation. Ancient Alien Theory proponents argue that extraterrestrial beings may have lent their advanced technology and expertise to help construct these remarkable edifices, providing solutions to seemingly insurmountable challenges like moving and positioning massive stone blocks with precision.

The Baghdad Battery is an intriguing artifact believed to date back to the Parthian or Sassanid periods, around 250 BCE to 650 CE. This artifact consists of a clay jar, a copper cylinder, and an iron rod, which some researchers suggest could have functioned as a simple galvanic cell, producing a small electric current when filled with an electrolytic solution. If this interpretation is accurate, it would imply that ancient civilizations possessed knowledge of electricity over a millennium before its discovery in the modern era. Ancient Alien Theory proponents argue that this advanced understanding of electricity could be the result of extraterrestrial intervention, with alien visitors sharing their technological knowledge with ancient humans.

The Antikythera mechanism is a fascinating ancient artifact discovered in a shipwreck off the coast of the Greek island of Antikythera in 1901.

This complex device, believed to date back to the 2nd century BCE, is considered the world's first known analog computer. The mechanism consists of a series of intricate gears and dials, designed to calculate and display astronomical information such as the positions of celestial bodies and the timing of eclipses. The sophistication and complexity of the Antikythera mechanism seem to defy the technological capabilities of its time. Ancient Alien Theory suggests that the knowledge and expertise required to create such an advanced device could have been provided by extraterrestrial beings, further supporting the notion of their involvement in the development of ancient technology.

By considering the possibility that extraterrestrial beings provided direct technological assistance to ancient human civilizations, Ancient Alien Theory offers a compelling alternative explanation for the remarkable achievements of our ancestors. The advanced artifacts and monumental structures that continue to perplex and inspire researchers today may be evidence of a shared history between humanity and otherworldly visitors, a collaboration that has shaped our understanding of the world and our place in the cosmos.

Modern UFO sightings and encounters have significantly contributed to the discussion of extraterrestrial spacecraft visiting Earth. These incidents often feature objects with advanced flight capabilities, such as rapid acceleration, hovering, and near-silent operation. While some of these sightings may be attributed to natural phenomena or misidentified human-made objects, others defy conventional explanations and provide compelling support for the existence of extraterrestrial spacecraft.

The Tic Tac UFO encounter, which occurred in 2004 during a military training exercise off the coast of Southern California, is one such incident that has garnered significant attention. U.S. Navy pilots reported observing an object shaped like a Tic Tac candy that exhibited remarkable flight capabilities, such as rapid acceleration, sharp maneuvering, and hovering. Subsequent video footage and radar data corroborated the pilots' accounts. Despite extensive analysis, no conventional explanation has been found for the observed phenomena, further fueling speculation about the existence of advanced extraterrestrial technology.

The Belgian UFO wave, which took place between 1989 and 1990, is another notable series of UFO sightings that lends credence to the possibility of extraterrestrial spacecraft.. During this period, thousands of people across Belgium reported observing triangular-shaped craft with lights at each corner. The objects were said to move silently and exhibit extraordinary flight capabilities. The Belgian Air Force investigated the sightings and even attempted to intercept the objects, without success. To date, the Belgian UFO wave remains unexplained and serves as a prominent example of potential extraterrestrial visitation.

The Rendlesham Forest incident, which occurred in December 1980 near RAF Woodbridge in Suffolk, England, is another compelling case in support of extraterrestrial spacecraft. U.S. Air Force personnel stationed at the base reported observing strange lights in the nearby forest and encountered an unidentified craft with hieroglyphic-like markings. The servicemen also experienced equipment malfunction and observed the craft emitting beams of light. Despite official investigations, no conventional explanation for the events has been provided, further supporting the idea of extraterrestrial involvement.

In 2006, a UFO sighting occurred at Chicago's O'Hare International Airport, one of the busiest airports in the world. Numerous witnesses, including airport employees and pilots, reported observing a disc-shaped object hovering above the airport before rapidly accelerating and disappearing into the sky, leaving a circular hole in the clouds. Despite the number of credible witnesses and the lack of a conventional explanation, the sighting has not been officially acknowledged, leading many to suspect a cover-up and reinforcing the possibility of extraterrestrial spacecraft visiting Earth.

These modern UFO sightings and encounters have played a crucial role in shaping public perception of extraterrestrial spacecraft and have spurred further investigation into the potential existence of advanced alien civilizations. The consistent reports of objects exhibiting remarkable flight capabilities and the inability to provide conventional explanations for these phenomena lend credibility to the idea that extraterrestrial spacecraft may be visiting Earth. As our understanding of the universe continues to grow, these sightings serve as a reminder that the possibility of extraterrestrial life and advanced technology cannot be entirely dismissed.

The Roswell incident, which took place in 1947, remains one of the most compelling and debated UFO events in history. The incident began when a rancher named W.W. "Mac" Brazel discovered unusual debris scattered across his property near Roswell, New Mexico. This discovery led to a series of events that would spark widespread speculation about extraterrestrial involvement and an alleged government cover-up.

Upon discovering the debris, Brazel reported his findings to the local sheriff, who in turn contacted the nearby Roswell Army Air Field. Military personnel were dispatched to the ranch to investigate and collect the wreckage. The Roswell Army Air Field initially issued a press release stating that they had recovered a "flying disc," which fueled rumors of a crashed extraterrestrial spacecraft. However, just a day later, the U.S. government retracted the statement, claiming that the debris was actually from a fallen weather balloon.

Many witnesses, including military personnel and civilians, have come forward over the years to provide testimonies that challenge the official weather balloon explanation. Some of these witnesses claim to have seen the wreckage firsthand and describe it as being composed of materials that were unlike anything they had encountered before, such as lightweight metal that could not be bent or broken. Others have reported seeing alien bodies recovered from the crash site, suggesting that the craft was not only of extraterrestrial origin but also carried passengers.

The U.S. government's handling of the Roswell incident has led to widespread accusations of a cover-up to conceal the truth about extraterrestrial visitation. Numerous documents and testimonies have emerged over the years, suggesting that the military engaged in a campaign to silence witnesses, confiscate evidence, and suppress information related to the crash. In 1994, the U.S. Air Force released a report stating that the Roswell wreckage was actually from a top-secret project known as Project Mogul, which involved high-altitude balloons carrying surveillance equipment to detect Soviet nuclear tests. However, many UFO researchers and Ancient Alien Theory proponents remain skeptical of this explanation, arguing that the evidence for extraterrestrial involvement is too compelling to dismiss.

The Roswell incident serves as a crucial piece of evidence for proponents of Ancient Alien Theory, who argue that it demonstrates a long-standing pattern of extraterrestrial visitation and government cover-ups. If the crash indeed involved an alien spacecraft, it would provide strong support for the idea that advanced extraterrestrial civilizations have been visiting Earth for centuries, potentially interacting with ancient human societies and influencing our history. The ongoing debate surrounding the Roswell incident underscores the importance of continued research and investigation into the possibility of extraterrestrial life and its potential impact on human civilization.

The Phoenix Lights event of 1997 remains one of the most significant and well-documented UFO sightings in modern history. Over the course of two nights, thousands of witnesses in Arizona and neighboring states reported seeing a series of strange lights and a massive V-shaped craft moving silently across the sky. The scale and duration of the event, along with the consistency of eyewitness accounts, have led many to believe that the Phoenix Lights are strong evidence of extraterrestrial visitation.

Witnesses of the Phoenix Lights described the V-shaped craft as being enormous, with some estimating its size to be larger than a football field. The craft reportedly emitted no sound as it moved through the sky, which contradicts the behavior of conventional aircraft. Additionally, the lights on the craft were arranged in a distinct pattern, with each light maintaining a consistent distance from the others. This level of organization and the unusual flight characteristics observed during the event have fueled speculation that the Phoenix Lights may have been a result of extraterrestrial technology.

Attempts to explain the Phoenix Lights as military flares or conventional aircraft have not satisfied many observers. While the U.S. Air Force claimed that the lights were simply flares dropped by A-10 Warthog aircraft during a training exercise, this explanation does not account for the silent, slow-moving V-shaped craft witnessed by many. Furthermore, the precision and duration of the lights' appearance are inconsistent with the behavior of military flares, which typically burn out and descend rapidly.

The lack of a satisfactory explanation for the Phoenix Lights has led many to consider the possibility of an extraterrestrial origin. Proponents of this theory argue that the advanced technology and flight capabilities exhibited by the V-shaped craft are indicative of a civilization far more technologically advanced than our own. The Phoenix Lights event, they contend, may represent a close encounter with an extraterrestrial spacecraft, providing further support for the existence of ancient aliens and their continued interest in Earth.

The Phoenix Lights incident, when considered alongside other unexplained UFO sightings and historical accounts of extraterrestrial contact, adds weight to the Ancient Alien Theory. If advanced extraterrestrial civilizations have visited Earth in the past, it is plausible that they continue to do so today. The Phoenix Lights event serves as a modern reminder of the potential for ongoing extraterrestrial involvement in human affairs and bolsters the argument for the existence of ancient aliens and their impact on human history.

The concept of extraterrestrial spacecraft and their potential interactions with human civilizations throughout history remains a fascinating and controversial topic. While much of the evidence supporting the existence of such craft is based on anecdotal accounts and ancient artifacts, the possibility of advanced extraterrestrial civilizations visiting Earth cannot be entirely dismissed. In fact, the recurring themes of otherworldly beings and advanced technology across various cultures and time periods lend credence to the idea that ancient aliens may have had a significant influence on human history.

As we continue to explore the depths of our universe and push the boundaries of our scientific understanding, the line between mythology and plausible scientific theories becomes increasingly blurred. Many myths and legends from diverse cultures contain stories of gods descending from the heavens in flying chariots or interacting with humans using advanced technology. These stories, when viewed through the lens of Ancient Alien Theory, may be interpreted as historical accounts of extraterrestrial visitations. By examining ancient texts, artwork, and artifacts, researchers can uncover potential clues that may help bridge the gap between mythology and scientifically grounded hypotheses regarding the existence of ancient aliens.

The study of purported extraterrestrial spacecraft, their designs, and technology can provide valuable insights into the potential capabilities of alien civilizations and inspire further research into advanced propulsion systems, energy generation, and materials science. These investigations encourage interdisciplinary collaboration between experts in fields such as archaeology, astronomy, physics, and engineering, fostering a more comprehensive understanding of our past and our place in the cosmos. The pursuit of knowledge regarding extraterrestrial spacecraft may also lead to breakthroughs in human technology, as we strive to emulate the advanced systems and materials that may have been utilized by ancient aliens.

As our understanding of the universe continues to grow, the search for evidence of extraterrestrial spacecraft and their potential impact on human history will undoubtedly remain an important and thought-provoking area of inquiry. This ongoing debate encourages critical thinking and fosters an open-minded approach to the study of our past, which is essential for uncovering the truth about the possible existence of ancient aliens. While conclusive evidence of extraterrestrial visitations may remain elusive, the exploration of this intriguing possibility has the potential to reshape our understanding of human history and inspire future generations to continue the search for answers beyond our own planet.

Chapter 19

The Future of Ancient Alien Theory

Area of Research, Creative Science Fiction or a Pathway to the Future?

The Ancient Alien Theory has made significant strides since its beginnings, captivating the interest of countless individuals and generating fervent discussions among experts and aficionados. As we continue to learn more about the universe and its mysteries, the theory evolves and adapts, carving out a space for innovative research and redefining the boundaries between science, history, and mythology. In this section, we delve into how the Ancient Alien Theory is transforming our pursuit of knowledge by unveiling new paths of exploration and questioning conventional beliefs, ultimately guiding us toward a more profound comprehension of our origins and our role in the cosmos.

The Ancient Alien Theory has contributed to a paradigm shift in our understanding of human origins and the development of ancient civilizations. By suggesting that extraterrestrial beings played a role in the progress of our ancestors, the theory challenges the traditional narrative and encourages us to reevaluate our assumptions about the capabilities of early humans. This shift in perspective paves the way for a more comprehensive and nuanced interpretation of our history, which encompasses not only the achievements of our ancestors but also the possible contributions of otherworldly visitors.

The far-reaching implications of the Ancient Alien Theory have fostered collaboration among various fields of study, including archaeology, anthropology, astronomy, and even theology. This interdisciplinary approach has facilitated a more holistic examination of the evidence and the development of new hypotheses that transcend the limitations of individual disciplines. By bridging the gaps between these fields, the theory fosters a richer and more inclusive understanding of our past, drawing on the diverse perspectives and expertise of researchers from different backgrounds.

The provocative nature of the Ancient Alien Theory has inspired researchers to adopt innovative research methods and technologies in their quest for evidence of extraterrestrial intervention. From the use of advanced imaging techniques like LIDAR and ground-penetrating radar to the application of artificial intelligence in deciphering ancient texts, the theory has spurred the development and utilization of cutting-edge tools that expand our ability to investigate the past. These advancements not only enhance our understanding of ancient civilizations but also contribute to the broader scientific endeavor, paving the way for new discoveries and breakthroughs across various fields.

The Ancient Alien Theory pushes the boundaries of our knowledge by questioning long-held beliefs and inviting us to consider alternative explanations for the mysteries of our past. This spirit of inquiry has led to a more open-minded and inclusive approach to the study of history, archaeology, and other related disciplines, fostering a willingness to entertain unconventional ideas and explore uncharted territory. As we continue to delve into the enigmatic corners of our past, the Ancient Alien

Theory serves as a reminder of the importance of curiosity, skepticism, and intellectual humility in our pursuit of truth and understanding.

Ultimately, the Ancient Alien Theory has the potential to unravel the cosmic tapestry that connects us to the universe at large. By positing that our ancestors had contact with extraterrestrial beings, the theory invites us to contemplate our place in the grand scheme of existence and ponder the possibility of other civilizations beyond our own. As we continue to explore the cosmos and unlock its secrets, the Ancient Alien Theory may hold the key to a deeper understanding of our shared heritage and our role in the unfolding story of the universe.

The discovery of exoplanets, or planets orbiting stars outside our solar system, has had a profound impact on the extraterrestrial hypothesis. With thousands of exoplanets identified and many more suspected, the possibility of habitable worlds with conditions suitable for life has become increasingly plausible. This greatly expands the potential for the existence of intelligent extraterrestrial life, which in turn lends credibility to the idea that ancient alien visitations may have occurred in our past.

Extremophiles, microorganisms capable of surviving in extreme environmental conditions, have been found thriving in some of Earth's most inhospitable environments. These discoveries have broadened our understanding of the types of environments in which life could potentially exist, suggesting that life may be more resilient and adaptable than previously thought. This research supports the Ancient Alien Theory by demonstrating that life could potentially flourish in a wide range of environments, increasing the likelihood that extraterrestrial life could have visited or even influenced the development of life on Earth.

Astrobiology, the interdisciplinary study of life in the universe, has made significant strides in recent years, further bolstering the Ancient Alien Theory. As researchers develop new methods and tools for detecting biosignatures, or signs of life, on other planets, the possibility of discovering extraterrestrial life becomes more feasible. This progress not only lends credibility to the idea of ancient alien visitations but also suggests that our understanding of life beyond Earth is constantly evolving, opening the door for new discoveries that could reshape our perceptions of human history and our place in the cosmos.

The growing body of scientific research supporting the extraterrestrial hypothesis has sparked renewed interest in examining ancient artifacts, texts, and structures through the lens of extraterrestrial intervention. Scholars are increasingly looking for evidence of advanced technology, knowledge, and capabilities that could suggest the presence of otherworldly beings in our past. This approach has led to reinterpretations of existing archaeological finds and prompted researchers to reevaluate the cultural and historical context of these artifacts, texts, and structures, with the potential to uncover new insights into our connections with extraterrestrial life.

As scientific research continues to lend support to the Ancient Alien Theory, it is likely that interdisciplinary collaboration between archaeologists, historians, astrobiologists, and other experts will become increasingly important in the pursuit of evidence for extraterrestrial involvement in human history. By combining their expertise, these researchers can develop new methodologies, uncover previously overlooked evidence, and challenge conventional wisdom in ways that may ultimately lead to a deeper understanding of our past and our relationship with the cosmos.

Göbekli Tepe, an archaeological site in Turkey dating back to the 10th millennium BCE, is considered one of the most significant discoveries in recent history. The site is composed of numerous stone circles with massive, T-shaped pillars adorned with intricate carvings of animals and abstract symbols. The sophistication of Göbekli Tepe's construction techniques, which predate the invention of metal tools and the wheel, has left researchers puzzled. Proponents of the Ancient Alien Theory argue that such advanced engineering and artistic skills could be indicative of extraterrestrial influence, suggesting that ancient astronauts may have provided knowledge or even direct assistance in the creation of this enigmatic site.

The Nazca Lines, a series of enormous geoglyphs etched into the desert landscape of southern Peru, have long been a source of fascination and debate among scholars and ancient alien enthusiasts. These massive figures, which include animals, plants,. and geometric shapes, can only be fully appreciated from a significant altitude, leading some to speculate that they were created as messages or markers for extraterrestrial visitors. The precision and scale of the Nazca Lines, coupled with their enigmatic purpose, have led proponents of the Ancient Alien Theory to argue that their creation may have been guided or inspired by otherworldly beings with advanced knowledge of mathematics, geometry, and aeronautics.

The Piri Reis map, a 16th-century Turkish cartographic masterpiece, has become a focal point for Ancient Alien Theory proponents due to its surprising accuracy and detail. The map, which predates the discovery of Antarctica by centuries, appears to show the continent's coastline free of ice, a feature that has led some researchers to suggest that its creator had access to knowledge beyond his time. Ancient Alien Theory supporters argue that this knowledge may have been passed down from extraterrestrial visitors, who could have shared advanced cartographic and geographical information with ancient civilizations, enabling them to create maps with remarkable accuracy.

The ancient site of Baalbek in modern-day Lebanon is home to some of the most massive and enigmatic stone structures known to archaeology. The enormous megaliths that make up the foundation of the site's temple complex, some weighing over 800 tons, have long puzzled researchers due to the apparent lack of technology and resources available to the ancient builders. Proponents of the Ancient Alien Theory argue that the construction of Baalbek would have required knowledge and capabilities beyond those of the civilizations known to have inhabited the region, suggesting that extraterrestrial beings may have played a role in its creation.

The ancient Inca fortress of Sacsayhuamán, located near Cusco, Peru, is renowned for its remarkable stone walls, composed of enormous, irregularly shaped blocks that fit together with extraordinary precision. The technique used to create these walls, known as polygonal masonry, has baffled researchers due to the apparent lack of advanced tools or engineering knowledge available to the Inca at the time. Ancient Alien Theory proponents propose that the construction of Sacsayhuamán may be evidence of extraterrestrial influence, as the precision and complexity of the stonework suggest advanced knowledge and techniques that could have been shared with the Inca by otherworldly visitors.

Developments in remote sensing technologies, such as LIDAR, have revolutionized the field of archaeology by enabling researchers to identify and study previously hidden or inaccessible archaeological sites. LIDAR, which uses laser pulses to create detailed topographic maps, has uncovered entire ancient cities and complex structures that were once obscured by vegetation or difficult terrain. The discovery of these sites has prompted some Ancient Alien Theory proponents to suggest that these newly uncovered sites may hold evidence of extraterrestrial contact or intervention. By continuing to utilize advanced imaging technologies, researchers may be able to uncover additional clues that support the existence of ancient aliens and their impact on human civilization.

Advances in materials science and nanotechnology have the potential to shed new light on the construction techniques and materials used by ancient civilizations. Some proponents of Ancient Alien Theory argue that certain artifacts, such as the Iron Pillar of Delhi or the Baghdad Battery, exhibit properties and levels of sophistication that seem beyond the capabilities of their time. By employing cutting-edge materials analysis techniques, researchers may be able to better understand the origins and properties of these mysterious artifacts, providing further support for the notion that ancient civilizations had access to advanced knowledge and technology, possibly through extraterrestrial intervention.

The use of robotics and autonomous systems in archaeological investigations is becoming increasingly common, providing researchers with new tools to explore and document ancient sites. These technologies can assist in navigating hazardous or hard-to-reach areas, such as underwater ruins or subterranean structures, and can help to uncover new evidence that may support the existence of ancient aliens. For example, robotic explorers have been used to investigate the interior of the Great Pyramid of Giza, revealing hidden chambers and passages that have eluded human researchers for centuries. The continued development and application of robotic technologies in archaeology could lead to the discovery of further evidence pointing towards extraterrestrial influences on ancient civilizations.

Artificial intelligence (AI) is increasingly being applied to the study of ancient technologies, with researchers using machine learning algorithms to analyze and interpret complex data sets, such as architectural designs, astronomical alignments, and construction techniques. By harnessing the power of AI, proponents of Ancient Alien Theory hope to uncover patterns and connections that may suggest extraterrestrial involvement in the development of ancient technologies. For example, AI-driven analysis of architectural plans or structural engineering techniques could reveal previously unnoticed similarities between disparate ancient sites, pointing towards the possibility of shared knowledge or influence from an advanced, non-human source.

The rapid advancements in technology have significant implications for the study and understanding of Ancient Alien Theory. Through the application of remote sensing technologies, materials science, robotics, and artificial intelligence, researchers are better equipped to investigate and uncover evidence that may support the idea that ancient civilizations had access to knowledge and resources beyond their own capabilities, possibly through extraterrestrial intervention. As technology continues to evolve, it is likely that new discoveries and insights will emerge, further challenging our understanding of human history and our place in the universe.

The application of artificial intelligence in the field of archaeology and linguistics has revolutionized the study of ancient texts. By using machine learning algorithms, researchers can analyze vast amounts of data, identify patterns, and decipher previously untranslated or misunderstood passages. This groundbreaking approach has the potential to reveal hidden layers of meaning within ancient texts, pointing to possible otherworldly influences in our history. In favor of the existence of ancient aliens, these AI-driven discoveries could provide further evidence of extraterrestrial involvement in shaping human culture and knowledge.

One of the most exciting applications of artificial intelligence in the study of ancient alien theory is the comparative analysis of myths and legends from diverse cultures. AI-powered algorithms can sift through vast databases of stories and folklore, identifying common themes, motifs, and narrative structures that transcend cultural and geographic boundaries. These shared elements, which may include accounts of celestial beings, advanced technology, and unexplained phenomena, could suggest a common history of contact with extraterrestrial civilizations.

Artificial intelligence can also be applied to the study of ancient art and iconography, helping researchers decode symbolism and identify patterns that may point to extraterrestrial influences. By analyzing images, sculptures, and other forms of artistic expression, AI algorithms can uncover recurring motifs and themes that may be indicative of encounters with advanced beings from other worlds. This innovative approach can provide valuable insights into the possible connections between ancient cultures and extraterrestrial visitors, lending support to the ancient alien theory.

Machine learning can play a vital role in reconstructing ancient technologies and understanding their origins. By analyzing the design and function of ancient artifacts, AI algorithms can help researchers reverse-engineer these technologies, offering insights into the advanced knowledge and techniques that may have been passed down by extraterrestrial visitors. This could lend further credence to the ancient alien theory, suggesting that our ancestors had access to otherworldly wisdom and resources.

As artificial intelligence continues to advance, its applications in the study of ancient alien theory are likely to expand and become even more sophisticated. Future developments in AI could lead to the discovery of previously unknown texts, artifacts, and archaeological sites, providing further evidence of extraterrestrial involvement in human history. By harnessing the power of AI, researchers and enthusiasts can continue to explore the possible connections between our ancestors and otherworldly beings, challenging conventional wisdom and shedding new light on our shared cosmic heritage.

The Search for Extraterrestrial Intelligence (SETI) and other similar initiatives have made significant strides in recent years, utilizing cutting-edge technology and innovative approaches to detect potential signals from advanced civilizations beyond our solar system. These efforts, which include radio telescope arrays, optical telescopes, and even the search for extraterrestrial artifacts, have the potential to revolutionize our understanding of the cosmos and lend credibility to the Ancient Alien Theory.

Astrobiology, the study of life in the universe, has become an integral part of SETI research. By understanding the conditions necessary for life and identifying potentially habitable exoplanets, astrobiologists contribute valuable insights that can guide SETI researchers in their search for extraterrestrial signals. This interdisciplinary approach strengthens the scientific foundation of SETI, making it more likely that the discovery of intelligent life beyond Earth could lend support to the Ancient Alien Theory.

If SETI or a similar initiative were to find definitive evidence of extraterrestrial intelligence, it could have profound implications for the Ancient Alien Theory. The discovery of an advanced civilization with the capability to travel between stars would lend credibility to the idea that such beings could have visited Earth in the distant past, influencing the course of human history. Moreover, understanding the motivations, technologies, and cultural practices of these extraterrestrial beings could provide valuable context for interpreting the ancient artifacts, texts, and structures that proponents of Ancient Alien Theory believe to be evidence of alien intervention.

One of the key challenges faced by SETI researchers is developing methods for interstellar communication that could facilitate the exchange of information between humans and extraterrestrial beings. By exploring various communication modalities, including radio signals, lasers, and even the transmission of physical artifacts, researchers are expanding the possibilities for detecting and understanding potential extraterrestrial messages. This line of inquiry not only holds promise for the future of SETI research but also has implications for the Ancient Alien Theory, as it raises the possibility that extraterrestrial visitors may have left behind encoded messages or artifacts that could be decoded and understood by future generations.

As SETI research continues to evolve and expand, it has the potential to influence the Ancient Alien Theory in a variety of ways. The development of new technologies and methodologies could lead to the discovery of previously undetectable signals or artifacts that lend support to the idea of ancient alien visitations. Additionally, collaborations between SETI researchers and experts in other fields, such as archaeology, linguistics, and anthropology, may yield fresh insights into the potential connections between Earth's history and extraterrestrial life. Ultimately, the future of SETI research and the Ancient Alien Theory are inextricably linked, with each offering new perspectives and possibilities for understanding our place in the cosmos.

The Ancient Alien Theory has sparked a renewed focus on the search for extraterrestrial life, driving scientists and researchers to look for evidence of past or present life on other celestial bodies. This interest has led to the development of advanced technologies and methods to detect biosignatures, the unique markers of life, in our solar system and beyond. As a result, space exploration missions are increasingly designed with the goal of uncovering signs of life, which, if found, could support the idea that our ancestors had contact with extraterrestrial beings.

The influence of the Ancient Alien Theory on space exploration is evident in the design and objectives of new space missions. For instance, NASA's upcoming Europa Clipper mission is set to explore Jupiter's moon Europa, which is believed to harbor a subsurface ocean that could potentially host life. Similarly, the James Webb Space Telescope, scheduled for launch in 2021, aims to study the atmospheres of exoplanets and search for potential biosignatures. These missions, inspired by the possibility of ancient alien contact, could provide crucial evidence to support or refute the Ancient Alien Theory.

The Ancient Alien Theory has also inspired researchers and engineers to develop new technologies for space travel, such as advanced propulsion systems and energy sources. These innovations could enable humans to journey deeper into space, allowing us to explore distant celestial bodies more efficiently and effectively. By studying the technologies and construction methods of ancient civilizations that may have had contact with extraterrestrials, we may unlock new insights into propulsion and energy systems, ultimately facilitating our own exploration of the cosmos.

As the Ancient Alien Theory gains traction, it encourages scientists to reevaluate our understanding of habitability and the conditions necessary for life to thrive. By considering the possibility that extraterrestrial life may have visited Earth in the past, we are prompted to expand our definition of habitable environments and investigate a broader range of celestial bodies for potential signs of life. This shift in perspective could lead to the discovery of previously overlooked life forms or even evidence of ancient alien activity.

The influence of the Ancient Alien Theory on space exploration has led to increased collaboration among scientists, researchers, and space agencies worldwide. The prospect of discovering evidence of extraterrestrial life or ancient alien contact has inspired a spirit of global cooperation, as nations work together to share resources, expertise, and technology in the pursuit of a common goal. This collaborative approach not only accelerates scientific progress but also fosters a sense of unity and shared purpose among the people of Earth, as we collectively seek to unravel the mysteries of our past and our place in the universe.

The Ancient Alien Theory offers a unique framework for understanding the universe that transcends the traditional boundaries of science and spirituality. By positing that extraterrestrial beings may have played a role in human history, the theory encourages us to consider a broader range of possibilities regarding the nature of reality and the origins of life. This openness to new ideas can help bridge the gap between scientific inquiry and spiritual exploration, fostering a more comprehensive and inclusive approach to the study of our place in the cosmos.

One of the key strengths of the Ancient Alien Theory is its ability to challenge dogma and promote open-mindedness in both science and spirituality. By proposing alternative explanations for phenomena that have traditionally been attributed to divine or supernatural forces, the theory invites us to reevaluate long-held beliefs and consider the possibility that our understanding of the universe may be more complex and interconnected than previously imagined. This process of questioning and reevaluation can help break down the barriers between scientific and spiritual perspectives, paving the way for a more integrated and holistic worldview.

The Ancient Alien Theory highlights the interplay between science and spirituality in the search for truth. By examining the potential connections between human history and extraterrestrial beings, the theory encourages us to recognize that both scientific evidence and spiritual insight can contribute to a deeper understanding of our origins and our place in the universe. By embracing this interdisciplinary approach, we can cultivate a more nuanced and inclusive perspective that transcends the limitations of traditional disciplinary boundaries.

The convergence of science and spirituality brought about by the Ancient Alien Theory has the potential to facilitate personal transformation. By encouraging individuals to question long-held beliefs and consider the possibility of a shared cosmic heritage, the theory can inspire a sense of awe and wonder that fosters spiritual growth and self-discovery. This process of personal transformation can ultimately lead to a greater appreciation for the interconnectedness of all things and a heightened sense of responsibility for our role in the unfolding story of the universe.

The Ancient Alien Theory's emphasis on our shared cosmic heritage has the potential to foster a sense of global unity and cooperation. By recognizing that we may all be connected through our interactions with extraterrestrial beings, we can begin to see ourselves as part of a larger cosmic family, transcending the divisions and conflicts that have traditionally defined our relationships with one another. This newfound sense of unity can inspire us to work together to address the challenges we face as a species and to create a more just, sustainable, and peaceful world for all.

The Ancient Alien Theory has the potential to serve as a catalyst for social change by encouraging people to rethink traditional paradigms and consider alternative explanations for the development of human civilization. By proposing that advanced extraterrestrial civilizations may have played a role in shaping human history, the theory challenges long-held beliefs about the origins of power, authority, and knowledge in our society. This shift in perspective could lead to a more open-minded approach to the study of our past, fostering greater intellectual curiosity and promoting interdisciplinary collaboration among researchers.

One of the most significant implications of the Ancient Alien Theory for social change is its potential to empower individuals to question established institutions. By suggesting that our ancestors may have been influenced by extraterrestrial beings, the theory raises questions about the foundations of religious, political, and educational institutions that have long claimed authority over our understanding of the past. This questioning could lead to a more critical examination of these institutions and their role in shaping our collective consciousness, ultimately promoting greater transparency and accountability.

As the Ancient Alien Theory gains traction, it has the potential to inspire people to seek alternative sources of wisdom and explore unconventional ideas about the origins and development of human civilization. By recognizing the potential influence of extraterrestrial beings on our history, individuals may be more open to considering other perspectives, from indigenous oral traditions to esoteric teachings, as valuable sources of knowledge. This openness could foster a greater appreciation for diverse cultural perspectives and encourage a more inclusive approach to the study of our past.

The Ancient Alien Theory can also serve as a catalyst for social change by fostering greater intellectual and spiritual freedom. By challenging conventional wisdom about the origins of human civilization, the theory encourages individuals to question their own beliefs and engage in a more open-minded exploration of the unknown. This process of questioning and exploration can lead to a deeper understanding of our place in the universe and our connection to other life forms, ultimately promoting greater intellectual curiosity and spiritual growth.

Lastly, the Ancient Alien Theory has the potential to inspire future generations to pursue careers in science, archaeology, and other fields related to the study of our past and the search for extraterrestrial life. By presenting human history as an interconnected web of influences, including possible interactions with otherworldly beings, the theory can spark the imagination of young minds and encourage them to explore new frontiers of knowledge. In this way, the Ancient Alien Theory can serve as a powerful catalyst for social change, inspiring a more open-minded and curious approach to our understanding of the past and our place in the cosmos.

Over the years, Ancient Alien Theory has steadily gained traction in popular culture, sparking curiosity and fascination among audiences worldwide. This phenomenon can be attributed to the engaging narratives and thought-provoking ideas presented by various media, which invite people to reconsider the origins of ancient civilizations, human achievements, and the possibility of extraterrestrial contact.

Literary works have played a crucial role in popularizing the Ancient Alien Theory, with authors like Erich von Däniken and Zecharia Sitchin presenting alternative interpretations of historical events, artifacts, and texts. Their groundbreaking books, such as "Chariots of the Gods?" and "The 12th Planet," have introduced millions of readers to the concept of ancient astronaut visitations and continue to inspire new generations of writers to explore the theory further.

Television shows like "Ancient Aliens" and documentaries such as "Out of the Blue" have brought the Ancient Alien Theory to a broader audience, presenting compelling evidence and expert testimonies in a highly accessible format. These productions have not only popularized the theory but also encouraged critical thinking and open-mindedness among viewers, prompting them to question established historical narratives and consider alternative explanations.

Blockbuster films like "Prometheus" and "Stargate" have incorporated elements of Ancient Alien Theory into their storylines, blending science fiction with historical speculation to create immersive and thought-provoking cinematic experiences. These films not only entertain audiences but also stimulate public interest in the potential connections between human history and extraterrestrial beings, fostering a growing fascination with the subject.

The influence of Ancient Alien Theory can also be seen in the realms of art, music, and fashion, with artists and designers drawing inspiration from the theory's central themes and iconography. From album covers and music videos to clothing and accessories, elements of the Ancient Alien Theory have permeated various aspects of contemporary culture, reflecting society's ongoing fascination with the enigmatic and the unknown.

Social media platforms and online communities have played an essential role in the dissemination and discussion of Ancient Alien Theory, connecting enthusiasts, researchers, and skeptics from across the globe. These digital spaces provide a forum for sharing ideas, evidence, and experiences, fostering a sense of community and collaboration among those who are passionate about exploring the potential connections between human history and extraterrestrial life.

As the Ancient Alien Theory continues to captivate the public's imagination, it is likely to inspire even more creative works that delve into the theory's many facets and implications. This ongoing evolution of the theory in popular culture will not only generate increased interest and awareness but also contribute to a deeper understanding of our past and our potential connections with otherworldly beings. In turn, this growing fascination with the Ancient Alien Theory may prompt further research, exploration, and inquiry, ultimately enriching our collective knowledge and expanding our horizons.

The future of the Ancient Alien Theory hinges on our ability to embrace interdisciplinary research that combines the efforts of archaeologists, historians, astrophysicists, and even psychologists. As we begin to view ancient civilizations and their accomplishments through the lens of extraterrestrial contact, we may uncover patterns and connections that were previously overlooked. This approach could yield groundbreaking discoveries, further legitimizing the Ancient Alien Theory and providing more substantial evidence to support the idea of ancient extraterrestrial intervention.

By fostering collaboration between researchers, enthusiasts, and even skeptics, the Ancient Alien Theory can continue to thrive and evolve. This collaborative exploration can lead to a more comprehensive understanding of our past and a more open-minded approach to the study of human history. Such cooperation can result in the development of novel methodologies and techniques for uncovering and interpreting evidence of ancient alien contact, pushing the boundaries of our knowledge and challenging conventional interpretations of our past.

The Ancient Alien Theory has the potential to inspire technological advancements that could reshape our understanding of the universe and our place within it. By contemplating the potential existence of advanced extraterrestrial civilizations, we can be motivated to push the limits of our own technological prowess. This drive could lead to breakthroughs in space travel, energy production, and communication systems, allowing us to unlock the mysteries of our cosmic neighbors and, perhaps, even establish contact with them.

As we continue to explore the implications of the Ancient Alien Theory, we may begin to reconsider our understanding of human potential and identity. The idea that our ancestors were influenced by extraterrestrial beings invites us to question our assumptions about human capabilities and the nature of intelligence. This reexamination of human potential could lead to a more profound understanding of our species, our history, and our capacity for growth and innovation.

Finally, the Ancient Alien Theory encourages us to cultivate a global perspective on human history, transcending cultural and geographic boundaries to consider the broader implications of our shared past. By examining the potential connections between ancient civilizations and extraterrestrial beings, we can foster a sense of unity and shared purpose that transcends national and ethnic divisions. This global perspective has the potential to inspire greater cooperation and understanding among the diverse peoples of the world, ultimately contributing to a more peaceful and harmonious future.

In conclusion, the Ancient Alien Theory holds great promise as a pathway to the future, offering new perspectives on human history, our place in the universe, and our potential as a species. By embracing interdisciplinary research, encouraging collaborative exploration, inspiring technological advancements, reconsidering human potential, and cultivating a global perspective on human history, this theory has the potential to reshape our understanding of ourselves and our cosmic connections, opening up new pathways of inquiry and challenging conventional wisdom. The Ancient Alien Theory will likely continue to be a thought-provoking and influential force as we search for answers about our origins and our place among the stars.

Chapter 20

Conclusion

The Impact and Significance of Ancient Alien Theory on Our World Today

The Ancient Alien Theory has captivated the imagination of millions and sparked heated debates among scholars, scientists, and enthusiasts alike. In this concluding chapter, we will explore the impact and significance of this theory on various aspects of modern society, including popular culture, scientific inquiry, and our understanding of human history. We will also consider the potential implications of the Ancient Alien Theory for our future.

The Ancient Alien Theory has had a remarkable impact on popular culture, permeating various forms of media and entertainment. Television series such as "Ancient Aliens" have garnered a dedicated following, drawing viewers in with fascinating discussions and analyses of potential evidence for extraterrestrial contact in our past. These shows have sparked widespread interest in ancient civilizations and their potential connections to alien visitors, inspiring people to delve deeper into history and question conventional explanations for ancient artifacts and events.

In the realm of cinema, movies like "Prometheus" and "Stargate" have captivated audiences with their imaginative portrayals of ancient alien encounters. These films have contributed to the popularization of the Ancient Alien Theory by weaving compelling narratives that showcase the possible influence of extraterrestrial beings on human history. By blending elements of science fiction and historical fact, these movies have encouraged viewers to consider the potential ramifications of ancient alien contact on our society and our understanding of the universe.

Countless books, both fiction and nonfiction, have also been inspired by the Ancient Alien Theory. These works have explored various aspects of the hypothesis, from detailed analyses of archaeological sites and artifacts to imaginative tales of ancient astronauts and their interactions with human civilizations. By presenting a diverse array of perspectives and interpretations, these books have enriched the discourse surrounding the Ancient Alien Theory and fueled further research into this captivating area of study.

Moreover, the influence of the Ancient Alien Theory extends beyond film and literature to other forms of media and entertainment, such as video games, comic books, and graphic novels. These creative outlets provide additional avenues for the exploration of human-extraterrestrial interactions, expanding the reach of the Ancient Alien Theory and engaging a wider audience.

The Ancient Alien Theory's impact on popular culture has led to a renewed interest in our history, fostering a sense of wonder and curiosity about our past and the potential role of extraterrestrial beings in shaping human civilization. This fascination has given rise to a new genre of speculative fiction that explores the "what ifs" of human-extraterrestrial interaction, presenting alternative explanations for well-known historical events and artifacts. Through this genre, people are encouraged to view the past through a new lens, questioning established beliefs about our origins and considering the possibility of a shared history with otherworldly beings.

The Great Pyramid of Giza, the oldest and largest of the three pyramids in the Giza complex, has long been a subject of fascination and intrigue. Constructed around 2560 BCE, this awe-inspiring monument has stood the test of time, and its precision engineering has led proponents of the Ancient Alien Theory to suggest that extraterrestrial intervention was involved in its creation.

According to this theory, the precise alignment of the pyramid with the cardinal points and its correlation with certain celestial bodies, such as the stars of Orion's Belt, indicate advanced astronomical knowledge that would have been beyond the capabilities of the ancient Egyptians. Additionally, the immense size and weight of the pyramid's blocks, some of which weigh over 50 tons, have led researchers to question how such massive stones could have been quarried, transported, and positioned with such accuracy without the assistance of advanced technology.

Puma Punku, an ancient site located in present-day Bolivia, is another example of a megalithic structure that has perplexed archaeologists and fueled the Ancient Alien Theory. This enigmatic complex, believed to have been constructed around 536-600 CE, features remarkably precise stonework and intricate carvings that suggest the use of advanced tools and technology.

The site's massive stone blocks, some weighing over 100 tons, are cut with such precision that they fit together seamlessly, without the need for mortar. The Ancient Alien Theory proposes that such exacting craftsmanship would have been impossible for the indigenous Tiwanaku culture to achieve without the aid of extraterrestrial technology. Moreover, the site's intricate carvings of seemingly abstract symbols have led some proponents of the theory to speculate that they may represent a form of communication with or instruction from extraterrestrial beings.

Baalbek, an ancient Phoenician city located in modern-day Lebanon, is home to some of the most massive stone blocks ever discovered. Known as the "Trilithon," these enormous stones weigh an estimated 800 to 1,200 tons each and have baffled researchers who struggle to explain how they were quarried, transported, and ultimately incorporated into the construction of the Temple of Jupiter.

The Ancient Alien Theory proposes that the technology required to move and manipulate such colossal stones could only have been provided by an advanced extraterrestrial civilization. This hypothesis is further supported by local legends that describe the site as being the landing place of the gods, which could be interpreted as ancient accounts of extraterrestrial visitors.

By challenging the conventional wisdom surrounding these enigmatic monuments, the Ancient Alien Theory has prompted researchers to reevaluate their understanding of the capabilities of ancient civilizations. In doing so, this theory has not only deepened our appreciation for the resourcefulness and ingenuity of our ancestors but has also fostered an ongoing search for alternative explanations that may shed light on the true origins and purpose of these awe-inspiring structures.

The Ancient Alien Theory offers a unique perspective on ancient texts and myths, suggesting that the seemingly fantastical elements within them may be based on actual encounters with extraterrestrial beings. This hypothesis provides a valuable framework for reinterpreting and analyzing these texts, revealing potential evidence of advanced technologies and knowledge that could not have been developed by ancient civilizations alone.

The Ancient Alien Theory also highlights the striking similarities between myths and legends across diverse cultures, suggesting that these stories may have a common origin in ancient alien contact. From the Greek myth of the Hyperboreans and the Hindu concept of Vimanas to the Native American stories of the Star People, these tales share a recurring theme of advanced beings visiting Earth and imparting wisdom to humankind. By exploring these connections, the Ancient Alien Theory sheds light on the possibility of shared extraterrestrial experiences that transcended cultural boundaries and shaped human history.

The influence of the Ancient Alien Theory extends beyond mythology and ancient texts, encompassing the study of architectural marvels from antiquity. Many researchers have drawn attention to the unexplained precision, scale, and complexity of structures like the Great Pyramid of Giza, the Nazca Lines, and the Moai statues of Easter Island. The Ancient Alien Theory proposes that these enigmatic creations may be the work of extraterrestrial beings who possessed advanced technologies and knowledge, thus providing a fresh perspective on the capabilities of ancient civilizations and the nature of their accomplishments.

By bridging the gap between science and mythology, the Ancient Alien Theory has inspired a growing interdisciplinary interest in the study of ancient civilizations, their beliefs, and their potential interactions with extraterrestrial life. This approach encourages scholars and researchers from various fields, including archaeology, anthropology, and astronomy, to collaborate and share their insights, leading to a more comprehensive understanding of our past and the potential influence of extraterrestrial beings on human history.

The Ancient Alien Theory's emphasis on the potential connections between ancient myths, texts, and extraterrestrial contact also has significant implications for the search for life beyond Earth. By considering the possibility that intelligent life forms may have visited our planet in the past, researchers are motivated to reevaluate the criteria for habitable worlds and the potential diversity of life in the cosmos. This expanded perspective has not only fueled the search for extraterrestrial life but has also enriched our understanding of the factors that contribute to the emergence and evolution of life in the universe.

The Ancient Alien Theory's ability to bridge the gap between science and mythology has led to numerous insights and discoveries in various fields of study. By reinterpreting ancient texts, uncovering cultural connections, and reevaluating architectural wonders, this theory has opened up new avenues of research and understanding in our quest to uncover the potential influence of extraterrestrial life on human history.

The Ancient Alien Theory has played a crucial role in advancing the field of astrobiology, which seeks to understand the origin, evolution, and distribution of life in the universe. By positing that extraterrestrial life has visited Earth in the past, the theory has stimulated interest in the search for other life forms beyond our planet. This curiosity has driven researchers to delve deeper into the study of the potential building blocks of life, such as amino acids, and the various chemical processes that could lead to the emergence of life on other celestial bodies.

The possibility of ancient alien contact has significantly impacted the search for habitable exoplanets. Scientists have been prompted to expand their criteria for what constitutes a habitable world, taking into account the diverse conditions that may support life. This shift in perspective has led to the discovery of an increasing number of exoplanets within the habitable zones of their host stars, as well as the identification of potentially habitable moons around gas giants. The Ancient Alien Theory has thus inspired a more inclusive approach to the search for extraterrestrial life, opening up new possibilities for the discovery of life-sustaining worlds.

The Ancient Alien Theory has also spurred the development of innovative technologies for space exploration. By suggesting that advanced propulsion systems and other means of traversing vast distances between stars may be possible, the theory has encouraged researchers to pursue groundbreaking methods of space travel. Concepts such as antimatter propulsion, nuclear fusion-powered spacecraft, and the Alcubierre warp drive are all being explored as potential means of achieving interstellar travel. These technological advancements not only bring us closer to the stars but also increase the likelihood of encountering extraterrestrial life in the future.

The Ancient Alien Theory has had a profound impact on the Search for Extraterrestrial Intelligence (SETI) initiative. By asserting that intelligent beings have visited Earth, the theory has reinforced the need for a systematic approach to the search for extraterrestrial life. SETI researchers have employed sophisticated methods, such as radio and optical telescope observations, to detect potential signals from advanced civilizations. The theory has also inspired the development of new search strategies, such as looking for signs of astroengineering projects or Dyson spheres, which could indicate the presence of technologically advanced extraterrestrial life.

Lastly, the Ancient Alien Theory has fostered a cosmic perspective on humanity's place in the universe. By suggesting that extraterrestrial life has interacted with Earth in the past, the theory has expanded our understanding of the interconnectedness of all life forms and the potential for future contact. This broader viewpoint has cultivated a sense of unity and shared purpose in the quest to unravel the mysteries of the cosmos and explore the potential for life beyond our planet. As we continue to reach for the stars, the Ancient Alien Theory remains a driving force in our pursuit of knowledge and understanding of extraterrestrial life.

The Ancient Alien Theory has had a transformative effect on religion and spirituality by challenging the conventional understanding of divine beings and the origins of human life. By proposing that extraterrestrial entities played a significant role in our past, this theory has inspired a reevaluation of religious texts, myths, and the very foundation of various belief systems.

The notion that gods described in religious traditions might be extraterrestrial beings who visited Earth in the distant past has led to the recognition of universal themes across multiple belief systems. This realization has deepened our understanding of the interconnectedness of human spirituality, transcending cultural and geographical boundaries. As proponents of the Ancient Alien Theory argue, this universality points to the possibility of a shared, otherworldly origin for many religious ideas.

The Ancient Alien Theory encourages a reinterpretation of religious texts, such as the Bible, the Qur'an, and the Bhagavad Gita, as potential accounts of ancient alien contact rather than purely divine revelations. For example, stories of divine beings descending from the sky, the biblical account of Ezekiel's vision of a "wheel within a wheel," and the descriptions of the Hindu gods traveling in Vimanas (flying vehicles) can all be viewed as possible evidence of extraterrestrial encounters.

The hypothesis that gods could be extraterrestrial beings has fostered a dialogue between religious scholars and proponents of the Ancient Alien Theory. This conversation has led to a deeper appreciation of the complexities and commonalities of human spirituality, transcending traditional dogma and opening the door for new perspectives on the nature of the divine.

For many individuals, the Ancient Alien Theory has had a profound impact on their personal spirituality and belief systems. By suggesting that our understanding of the divine may be rooted in encounters with extraterrestrial beings, this theory has challenged long-held beliefs and encouraged individuals to explore alternative explanations for their spiritual experiences. This reevaluation has led to the development of new spiritual practices and beliefs that incorporate the idea of ancient alien contact, creating a fusion of traditional religious thought and modern speculation about the nature of the cosmos.

The Ancient Alien Theory's impact on religion and spirituality is multifaceted, encompassing the universality of religious themes, reinterpretation of sacred texts, dialogue between religious scholars and proponents, and the transformation of personal belief systems. By offering an alternative perspective on the nature of gods and the origins of human spirituality, the Ancient Alien Theory has fostered a deeper understanding of the complexities and commonalities of our spiritual heritage, while inspiring the exploration of new possibilities in the realm of the divine.

As our understanding of the universe continues to expand, the Ancient Alien Theory is poised to play an increasingly significant role in shaping our worldview. With each new exoplanet discovery and advancement in astrobiology, the possibility of encountering extraterrestrial life becomes more plausible. This growing likelihood has the potential to reinvigorate interest in the Ancient Alien Theory and its implications for our past, present, and future.

One area where the Ancient Alien Theory can have a significant impact on our future is in the realm of interstellar travel and communication. The theory's premise that ancient civilizations were visited by extraterrestrial beings suggests that travel between stars is not only possible but has already occurred in our past. This idea drives researchers to explore innovative methods of space propulsion and communication, potentially leading to groundbreaking discoveries that could bring humanity closer to achieving interstellar travel.

Another compelling aspect of the Ancient Alien Theory is its capacity to inspire wonder and curiosity about the achievements of ancient civilizations. By suggesting that our ancestors may have benefited from extraterrestrial knowledge and technology, the theory encourages a more inclusive and global perspective on human history. This outlook can lead to increased collaboration among researchers from different disciplines and cultures, fostering a more holistic understanding of our past and its implications for our present and future.

The Ancient Alien Theory has also revitalized interest in the Search for Extraterrestrial Intelligence (SETI). By positing that intelligent life exists beyond Earth and has interacted with humanity in the past, the theory lends credence to the possibility that we may one day detect signals from other advanced civilizations. This renewed enthusiasm for SETI has the potential to stimulate funding and research into advanced technologies and strategies for detecting extraterrestrial signals, which could ultimately lead to the discovery of other intelligent life forms in the cosmos.

Finally, the Ancient Alien Theory has the power to awaken our sense of cosmic interconnectedness and encourage a spirit of exploration and discovery. As we continue to probe the depths of space and unlock the mysteries of our own history, the Ancient Alien Theory will undoubtedly remain a thought-provoking force in our quest for answers about our origins and our place in the cosmos. By challenging conventional wisdom and pushing the boundaries of scientific inquiry, the theory can inspire new philosophical debates on the nature of human identity and our relationship with the universe.

The Ancient Alien Theory has the potential to reshape our understanding of the cosmos and our place within it. By inspiring new approaches to interstellar travel, fostering a global perspective on ancient civilizations, promoting the search for extraterrestrial intelligence, and influencing the philosophy of science and human identity, the theory promises to play a critical role in guiding humanity's ongoing quest for knowledge and understanding. As we continue to explore the universe and delve deeper into our own history, the Ancient Alien Theory will remain an influential and thought-provoking force, driving us to seek answers to the most profound questions about our existence and our place among the stars.

In conclusion, the Ancient Alien Theory has played a significant role in shaping our understanding of human history and our place in the universe. Its impact on various facets of society, including popular culture, scientific inquiry, and religious beliefs, has been profound and far-reaching. By challenging conventional wisdom and bridging the gap between science and mythology, this theory has expanded our knowledge of the past and opened up new possibilities for our future.

The Ancient Alien Theory has encouraged us to reevaluate our assumptions about the capabilities of ancient civilizations and the nature of our historical narrative. Through the lens of this theory, enigmatic artifacts, structures, and myths take on new significance, pointing towards the possibility of extraterrestrial involvement in our past. By considering the existence of ancient aliens, we gain a deeper appreciation for the ingenuity of our ancestors and the complexities of their societies.

Furthermore, the Ancient Alien Theory has fostered a sense of cosmic interconnectedness and curiosity that drives our exploration of the universe. By suggesting that intelligent beings have visited our planet in the past, this theory has fueled the search for extraterrestrial life and encouraged scientific research into advanced propulsion systems and potential means of interstellar travel. As our knowledge of the cosmos grows, the Ancient Alien Theory serves as a powerful reminder of the potential for discovering other life forms and the importance of understanding our place in the universe.

Lastly, the Ancient Alien Theory has had a profound impact on religion and spirituality, prompting a reevaluation of traditional beliefs about the nature of gods and the origins of human life. By proposing that extraterrestrial beings may have played a role in our past, this theory has inspired a dialogue between religious scholars and proponents of the Ancient Alien Theory. This dialogue has led to a deeper appreciation of the complexities and commonalities of human spirituality and the possibility of a shared, otherworldly origin.

As we continue to explore the universe and unravel the mysteries of human history, the Ancient Alien Theory will remain a vital and influential component of our ongoing search for truth and meaning in the cosmos. By keeping our minds open to the possibility of ancient aliens, we allow ourselves to entertain alternative explanations and embrace the spirit of discovery that has driven human progress for millennia. In doing so, we can continue to expand our understanding of the past and chart new courses for our future.

About the Author

William M. Rowe

William M. Rowe is a passionate author who has dedicated over 50 years to researching and gathering information about the possibility of extraterrestrial life and its potential impact on human civilization. His book, "Ancient Alien Theory: A Comprehensive Guide to the Evidence," covers a wide range of topics related to ancient alien theory, spanning from theories of the ancient astronaut hypothesis to the search for extraterrestrial communication.

Rowe's research delves deep into the evidence of ancient alien contact, including ancient artifacts and anomalies, stories and legends, and the evidence found in ancient texts. He also explores the role of governments and science in the cover-up of extraterrestrial life and the psychological effects and impacts of alien abduction and close encounter experiences.

Rowe also covers a range of scientific theories related to extraterrestrial life, such as theories of intergalactic travel, planetary engineering, and terraforming. His work also delves into the technology and designs of extraterrestrial spacecraft.

The book concludes with a discussion of the impact and significance of ancient alien theory on our world today and its potential implications for our future.

Overall, William M. Rowe's extensive research and knowledge have resulted in a comprehensive guide to ancient alien theory that presents a thought-provoking argument for the possibility of extraterrestrial life and its potential impact on our world.

Printed in Poland
by Amazon Fulfillment
Poland Sp. z o.o., Wrocław

53504962R00223